ARRL's
Vertical Antenna Classics

Compiled by Bob Schetgen, KU7G

Production Staff:

Robin M. Micket
Jodi Morin, KA1JPA
Joe Shea
Paul Lappen

Cover:

Design: Sue Fagan

Published by:
The American Radio Relay League
225 Main Street
Newington, CT 06111-1494

This work is Publication no. 201 of the Radio Amateur's Library, published by the ARRL.

Printed in USA

First Edition
Third Printing

ISBN: 0-87259-521-8

Foreword

Vertical antennas are all around us. AM broadcasting stations use them. So do cellular telephones and a lot of radio devices in between. Vertical monopoles range from slender spires stretching skyward to stubby rubber ducks poking out of pockets.

Common as they are, often they can be made to work better. Would you like your Amateur Radio station to "speak" with more authority? Better antennas are often the easiest and most cost effective approach. The projects and tips in this book can help you.

Computer modeling is a big part of modern antenna work, and this book includes a good foundation on which to build. Look at the "Beginner's Guide..." and "... Other Edge of the Sword" articles to get you started and map the pitfalls. The modeling techniques are useful for *all* antennas—not just verticals!

How about directional antennas? Arrays of two or more verticals can do the job. Vertical arrays have several advantages over tower mounted arrays: less visual impact, less damage from weather, the ability to change directions instantly, and there's no tower to climb!

You will also find information about "loading" verticals to reduce their size and several discussions of the radials and ground systems that are so important to vertical antenna performance.

We've pulled together the best information from many ARRL publications in this convenient guide to vertical antennas. You will also want to refer to *The ARRL Antenna Book* for fundamental antenna theory and RF-safety information.

David Sumner, K1ZZ
Executive Vice President

Newington, Connecticut
September 1995

Contents

Chapter 1: Theory and Modeling

Chapter 2: VHF and UHF

Chapter 3: HF

Chapter 4: Directional Arrays

Chapter 5: Reduced Size

Chapter 6: Radials and Ground Systems

Chapter 1

Theory and Modeling

from September 1978 *QST*

Designing a Vertical Antenna

Graphs cut through the mathematical headaches of antenna design. Put them to work and build a vertical that will shake the air with energy.

By Walter Schulz,* K3OQF

Here is a vertical designed and built from graphs contained in *The ARRL Antenna Book* and *ARRL electronics data book*. In my case the antenna was completely made from discarded Yagi beam elements — a junk box vertical!

By combining information found on transmission lines and antennas in *The ARRL Antenna Book* a design concept may be realized. Explaining further, antennas go through impedance variations in a manner similar to transmission lines. An open-ended transmission line exhibits inductive and capacitive reactances above and below "resonance," respectively. However, at resonance inductive reactance cancels capacitive reactance, leaving only a resistive component. The characteristics of a vertical are similar to those of an open ended transmission line.[1] Engineers use this concept to calculate conjugate impedance at an antenna feed point.

By using graphs of the universal reactance curves[2] and radiation resistance curves,[3] knowledge of mathematics other than simple arithmetic is not necessary. These charts make the solution to feed-point conjugate impedance and top loading problems simple.

Let's Design a Vertical

The antenna selected for illustration in this article is a top-loaded vertical for the 40-meter band, operating at one quarter wavelength or 90 electrical degrees.

Electrical degrees are often employed as units of measure when working with antennas. Their use not only helps one to mentally visualize antenna length, regardless of wavelength, but they are essential when working with the graphs mentioned above.

* PO Box 52056, Bustleton Station, Philadelphia, PA 19115

In the broadcast industry the practical physical limit for top loading is considered as approximately 30 electrical degrees[4] when applied to a disk. To find the actual physical length of a vertical antenna having this full limit of top loading, subtract 30° from 90°. The resulting 60° may then be converted to feet (or meters) by this equation:[5]

$$\text{Length in ft} = \frac{2.73 \times \ell}{f_{MHz}} \qquad \text{(Eq. 1a)}$$

$$\text{Length in m} = \frac{0.83 \times \ell}{f_{MHz}}$$

where ℓ = length in degrees

Thus,

$$\text{length} = \frac{2.73 \times 60}{7} = 23.4 \text{ ft} \qquad \text{(Eq. 1b)}$$

In order to proceed to the next step in the calculations, one should survey the aluminum stock on hand, and select masting having the desired outside diameter (OD). The tubing selected as an example for this article had an outside diameter of one inch. To obtain dimensions in meters (millimeters) multiply feet by 0.3048 (304.8) inches by 0.0254 (25.4).

Fig. 1 — Universal reactance curves for open and shorted transmission lines.

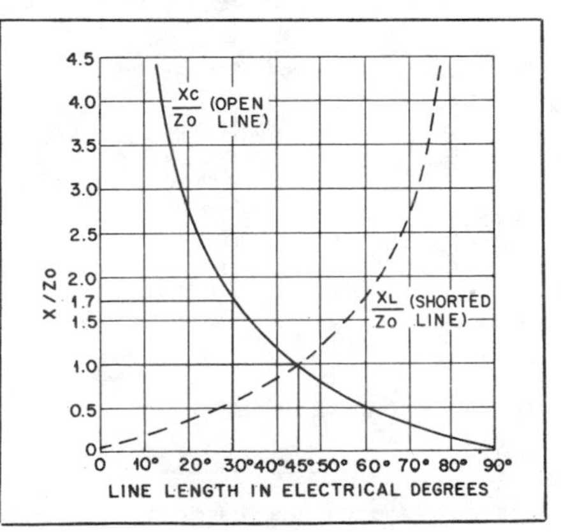

Let's now consider the vertical mast as an open-ended transmission line, so that the conjugate impedance and 30° top-loading dimensions can be determined. This equation is for computing the characteristic impedance:[6]

$$Zo = 60 \left[\ln \left(\frac{2h}{a} \right) - 1 \right] \qquad \text{(Eq. 2)}$$

where

ln = natural log (2.3 times the common log),

h = length or height of vertical mast in inches

a = *radius* of mast in inches

Thus,

$$Zo = 60 \left[\ln \left(\frac{2 \times 280.8}{0.5} \right) - 1 \right]$$

$$= 361 \text{ ohms}$$

By referring to the universal reactance curves in Fig. 1 the 30° of top-loading reactance can be found. Look across on the abscissa (line length in electrical degrees) finding 30°, and run along the projection vertically to a point on the Xc/Zo (open line) curve. At that point proceed horizontally toward the ordinate reading X/Zo = 1.7. By transposing X/Zo = 1.7 we observe that X = 1.7 × Zo, with the result X = 1.7 × 361 = 614 ohms reactance for 30° top loading.

How to Find Your Hat Size

Refer to Fig. 2 for the nomograph for LC constants, taken from the *ARRL data book*.[7] Place a ruler across 7 MHz and 614 ohms Xc reactance. The ruler crosses the capacitance line at 37 pF. For 30° of top loading, 37 pF of capacitance is required.

Turn next to Fig. 3, the graph of capacitance vs. diameter,[8] where the proper diameter for 37 pF can be found. Note

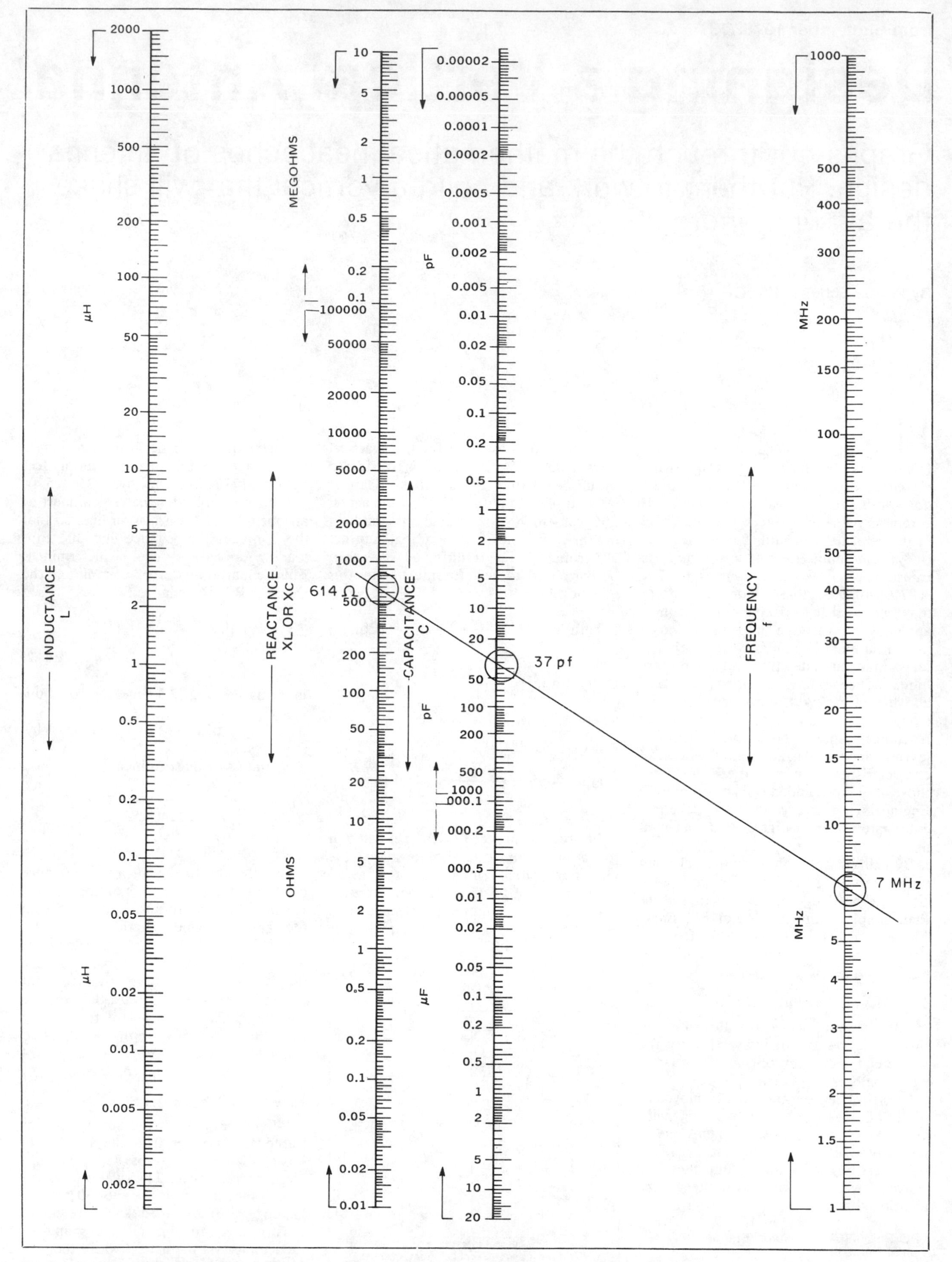

Fig. 2 — Nomograph for LC constants showing how values for the antenna described in the text are plotted.

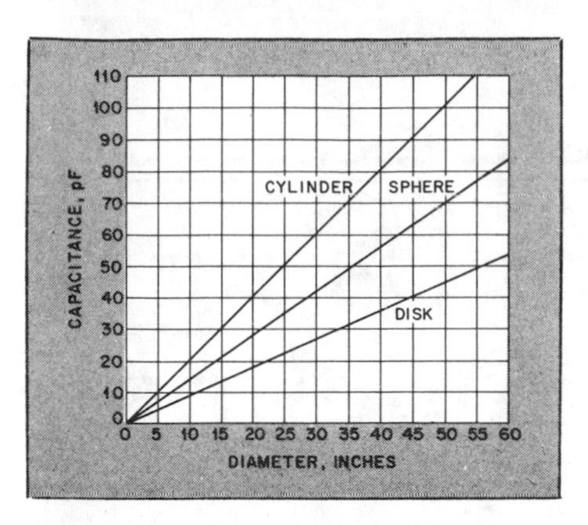

Fig. 3 — Capacitance of sphere, disk and cylinder as a function of diameter. The cylinder length is assumed equal to its diameter.

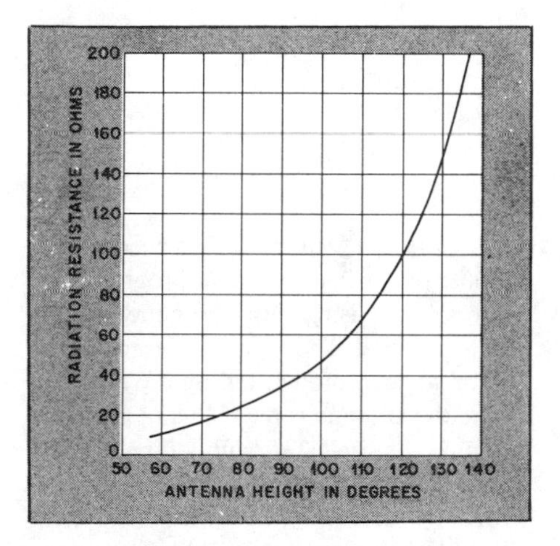

Fig. 4 — Radiation resistance vs. antenna height in degrees, for a vertical antenna over perfectly conducting ground or a highly conducting groundplane.

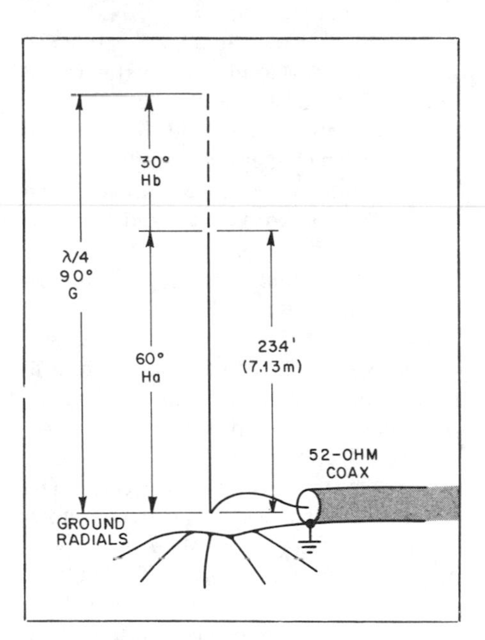

Fig. 5 — Dimensions for a quarter-wave vertical antenna with 30° of top loading. The dimensions in electrical degrees are provided. H_a represents the vertical portion and H_b is the capacitance hat. The antenna is series fed by the coaxial transmission line. There are 60 radials, each 0.2 wavelength long, in the ground system.

the position of 37 pF on the ordinate and the position of the point marked "disk" on horizontal projection. At this point follow the projection down to the abscissa (diameter, inches). The value, 40 inches, is the required diameter of the top-hat disk.

The skeleton disk shown in the photograph is fashioned into a wagon-wheel configuration. Six 20-inch lengths of 1/2-inch wide OD aluminum tubing are used as spokes, each emanating from the hub at equidistant intervals. The spokes terminate at a loop made of no. 14 copper wire. Note that the loop will increase the capacitance slightly.

To find conjugate impedance refer to the radiation-resistance-vs.-antenna-height graph, Fig. 4. Looking at the curve we see that for 90° (on the abscissa) we will have 36-ohms radiation resistance (on the ordinate). An estimated radial ground system loss resistance of 4 ohms for 60 radials, each 0.2 wavelength long,[9] may be added to the 36-ohms radiation-resistance value. This results in a total resistive value of 40 ohms. (Note: 60 radials were used with the antenna selected for the example). [A more complete discussion of top-loaded verticals can be found starting on p 6-17 in the 18th Edition of the *ARRL Antenna Book*. – Ed.]

Again referring to the universal reactance curves, Fig. 1, we see that 90° on the abscissa yields a reactance value of zero. Therefore, the conjugate impedance at the feedpoint is $Z = 40 \pm j0\,\Omega$. The electrical design for the completed antenna is shown in Fig. 5.

A further word about the universal reactance curves; these curves in reality are trigonometric functions. The two functions of interest here are $X/Zo = \cot \theta°$ for open transmission lines and $X/Zo = \tan \theta°$ for shorted lines. Knowing this information one could make his own graph using trigonometric tables.

For beginning radio amateurs without knowledge of the Smith Chart, use of the graphs facilitates vertical antenna design. They offer numerous possibilities in planning with a simple and direct approach.

When the 40-meter antenna was finally constructed, stations in Europe could be worked on a daily basis barefoot from the Philadelphia area. On several occasions stations as far away as the Indian Ocean have been worked.

Footnotes

[1]Jordan, *Electromagnetic Waves and Radiating Systems,* Prentice-Hall, Inc., 1968. pp. 388-396.
[2]*The ARRL Antenna Book,* 1968, p. 80.
[3]Fig. 2-74, *The ARRL Antenna Book,* 1974, p. 60.
[4]Laport, *Radio Antenna Engineering,* McGraw-Hill, Inc., 1952, p. 80.
[5]Department of Navy, *Naval Shore Electronics Criteria: HF Radio Antenna Systems,* Naval Electronic Systems Command, Washington, DC, 1970, p. A-6.
[6]Jasik, *Antenna Engineering Handbook,* McGraw-Hill, Inc., 1961, p. 19-3.
[7]*ARRL electronics data book,* 1976, p. 27.
[8]Fig. 2-80, *The ARRL Antenna Book,* 1974, p. 62.
[9]Stanley, "Optimum Ground Systems for Vertical Antennas," *QST,* December, 1976, pp. 13-15.

Joseph Blair, W2UI, stands beside a top-loaded 40-meter vertical antenna that is the key to regular contacts with stations in Europe.

A close view of the capacity hat for a 40-meter vertical antenna. The radial arms terminate in a loop of copper wire.

A Beginner's Guide to Using Computer Antenna Modeling Programs

By L. B. Cebik, W4RNL
1434 High Mesa Dr
Knoxville, TN 37938-4443

When MININEC became available to antenna experimenters as an antenna modeling computer program, they absorbed it with relish. It saved them hours, if not days, of futile construction effort on designs that would not improve performance. Now the program is available in at least three versions for the IBM PC and compatible computers (MININEC3, MN, and ELNEC[1]) to the average ham at reasonable costs. Depending upon the version, it will run on PCs with or without math coprocessors. Two versions (MN and ELNEC) produce excellent screen graphics of antenna patterns, along with documentation of the design, source impedance factors, and current distribution.

Are these programs really useful to the beginning and moderately experienced ham? The answer is a resounding YES! When used within their limitations, these programs can go far beyond textbooks in teaching us why our antennas act the way they do.[2] They can also help us make better decisions on what antennas to build or buy and how to mount them. However, after a brief period of sampling the test designs included with the program, the antenna modeling program may end up in a disk file box. The reason for discarding these valuable programs is that most of us fail to understand all they can tell us, and that—in turn—is because we do not set up procedures to squeeze meaningful information out of the program.

The purpose of this article is to show the beginner how to start using an antenna modeling program effectively. A good beginning requires three areas of effort: (1) setting up certain program basics, (2) setting up consistent modeling conventions, and (3) developing a baseline of information about basic antennas located on one's own property. The first step permits us to focus on and master the essentials of the program, saving advanced features for later. Step 2 allows us to model accurately and confidently, with minimal error in comparing one design with another. Step 3 allows us to interpret intelligently the patterns that emerge from new designs we try. Once we have mastered both the program and what it can tell us, we can expand our knowledge by using its advanced features.

The suggestions presented here are no substitute for mastering the instructions that come with the program; instead, they are designed to supplement those instructions. The aim here is to make the program and its procedures as useful and instructive as possible, even for the beginning antenna modeler. The program I use is ELNEC, but the steps suggested here can be translated for use with nearly any MININEC-based program.

Setting Up the Program

Advanced users of ELNEC and similar programs require considerable flexibility, so the programs offer many options to the user. Often these options inhibit the beginning student of antennas by offering choices among which the user cannot decide. Therefore, the first step in getting the most out of the antenna modeling program is to make decisions, even if they are initially made for weak reasons. For the new user, convenience may be the best reason available.

ELNEC offers a menu with many options, only a few of which we need for the purpose of getting used to the program. The "Wires" entry is for describing the antenna we shall model. The "Sources" entry is also crucial to each model, telling the program where the antenna is fed. Most of the basic antennas we shall start with—dipoles, verticals, Yagis, and the like—use only one source. Until we start modeling trap dipoles and beams, we shall likely have no use for the "Loads" entry (resistances and reactances as part of the antenna), so we may leave it at 0. Other fields to leave at default initially are the "Analysis Resolution" entry (1°), the "Step Size" entry (5°), and the "Field(s) to Plot" entry (total field only). After we become more advanced or develop special interests in the program, we may wish to alter those entries.

Two entries where we must make an initial decision are "Units" and "Ground Type." For US hams, feet and inches are the most commonly encountered antenna measurements. For HF work, start and stick with measurements in feet. This will ease the problem of calculating specific dimensions of antennas by sending us through the same steps with each antenna.

The selection of the type of ground to use is a bit more complex to decide. The program offers free space, perfect ground, and real-ground choices. My personal preference is for real ground. The default real-ground description uses average soil (as explained in the manual). Unless there is good reason to change this entry, initially stick with it. However, if you know the electrical characteristics of your soil and surrounding terrain (and water bodies, if present), it may pay to go through the setup procedures in the manual for establishing a detailed ground description. Remember, however, that the program does not account for yard clutter that may affect an antenna. Later you can compare real-ground patterns with patterns over perfect ground to see what differing soil conditions can do to an antenna's gain and pattern. You may also wish to compare your real-ground patterns with the free-space patterns so common in textbooks. Those exercises will be enlightening, but for comparing one antenna with another, stay with a consistent description of the ground. There are program limitations that the selection of real ground cannot overcome; we shall note the most significant one for the beginning modeler later.

Only two more menu entries require comment at this point. For the antenna "Title," choose your words and abbreviations carefully to pack in as much information as possible. That will allow you to easily distinguish one antenna design from another. "Dipole" is not a very good choice for a title, but "10 M Wire Dipole, 30 Ft Up" might be. The "Frequency" entry also requires care. For each band on which you compare different antennas, use the same frequency or set of frequencies. Do not shift from one end of a band to the other when changing antenna designs (unless you have a special reason for doing so); your comparisons may not be valid. If necessary, evaluate an antenna design at two or three frequencies within a band, using the same frequencies for all antenna designs.

Having now decided most of the menu entries in advance, we have reduced the remaining entries to a manageable few. To display and print out our results, we shall deal with decisions involving "Plot Type" and "Azimuth/Elevation Angle" later on. First we have to model our antenna design in order to enter it into the computer, and that takes some special forethought.

Setting Up Modeling Conventions

Getting good results from an antenna program begins with pencil and paper. Whether you are modeling your own antenna, an idea from a handbook, or an experimental design, you will have to put the figures on paper and change their form to what the program wants to see. Therefore, you will want to develop a standard notebook page for each antenna you model. Here are the preliminary items that should go on the page.

Item 1

Make a neat sketch of the antenna, including all dimensions. This includes any changes in element dimensions, a common occurrence for beams using aluminum tubing.

Item 2

Tabulate the details of each element, including the total length, the length of each piece or section of wire or tubing that makes up the element, and their diameters or wire sizes. Also include the spacing between elements for multielement antennas. At this point, it may be good to place the tabulated data in a column on the left side of the notebook page, since you will want to manipulate that data before entering it into the program.

All MININEC programs require entry of the data in the form of x, y, and z coordinates for the ends of each "wire" or element section, where x is the standard left-right axis, y is the standard front-back (on the computer screen, up-down) axis, and z is the height of the element above ground. Each element section made from a different diameter wire or tube has two entries, corresponding to the two ends of the section. It makes no difference to the program whether you model the antenna using x for the element length and y for the spacing between elements or vice versa. The convention you choose depends on which system is most convenient.

Since the chief problem I have in setting up models for the computer is entry errors, I have chosen x for the antenna element lengths and y for the spacing. This puts the coordinates most likely to be numerous or to change (values for x) in the left-hand column of the entry readout, where I can survey them easily for errors. And hard-to-see errors do occur, as when I forgot a decimal point and made part of an antenna element 75 inches in diameter rather than 0.75 inches. Adopting this convention will require that we take our elevation plots at an angle of 90° rather than at 0°, as noted later on.

Let's go a step further. Although you can enter any set of x and y coordinates, so long as the element sections line up, you will probably make fewer errors if you set up your antenna symmetrically around x = 0 and y = 0. Since y is the antenna element spacing dimension, set the driven element at y = 0. A simple dipole, of course, will have no other element. A Yagi might have either or both a director and a reflector. Set the reflector behind the driven element by giving it a negative y sign. A reflector 6 feet behind a driven element shows up as y = –6 on the chart. Any directors receive positive spacing. This convention allows you to identify elements in your setup chart.

Set up your antenna elements symmetrically around x = 0. For a single-wire element, this means taking the total length and dividing it by two. The ends of the wire then have identical x entries, but one is negative while the other is positive. For example, a 66-ft dipole would have x entries of 33 and –33 for its two ends. Multisection elements of different diameters are only slightly more complex. I have a portable 10-meter dipole made from 4-foot sections of ¼-inch diameter aluminum rod with end pieces of 3⁄16-inch rod. The overall length is 16.6 feet. We can consider the center rods as one 8-foot piece, which gives us –4 and +4 for the pair of x coordinates. Each end piece is 4.3 feet long. The x coordinates for the outside ends will be –8.3 and +8.3, with the other ends of the end rods having the same x coordinates as the center rod.

There are two advantages to using this system. First, you can spot errors more easily on the element entry screen or printout. Just look for numbers that are supposed to be the same except for a sign change. (And be sure to survey all the signs for correct negative or positive values while you are at it.) Your notebook page can have compact entries in preparation for computer work.

Item 3

Enter in columns next to each element section the arithmetic you did to determine element coordinates. For example, the end element sections of the dipole might show 16.6 ft – 8 ft = 8.6/2 = 4.3 ft ±4 ft = ±8.3 ft.

Item 4

Enter the final x coordinates for each element section. If you are careful, you can use abbreviated notations. You can put sections 1 and 3 together as ±8.3 and 4.0, with section 2 as ±4.0. This means that section 1 goes from –8.3 to –4.0, section 2 from –4.0 to +4.0, and section 3 from 4.0 to 8.3. If you are not comfortable with abbreviated listings, then list each piece or section separately. Note that what we have called "pieces" or "sections" are called "wires" in the program. What the program calls "wires" may actually be wire conductors, or they may be lengths of metal tubing or rod.

Figs 1 and 2 illustrate two notebook pages, one for the simple dipole, the other for a 3-element Yagi. Notice that all the entries are in feet and decimal parts of feet. Therefore, add another item to the pencil and paper you need in order to plan your entries: a calculator. You will often encounter dimensions like 9 feet 9½ inches, which a calculator easily converts to 9.7917 feet.

Band	Antenna Type	# El.	Height	Ground	Date	Run ?
10 M	Dipole (portable)	1	20'	R	01 / 12 / 1991	x

L. B. Cebik, W4RNL

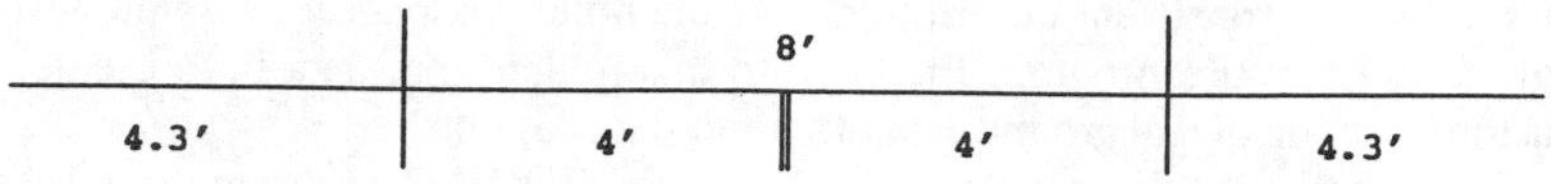

8'

4.3' 4' 4' 4.3'

El. #	Lgth feet	Sp to DrEl	Calculations	Wire #	E1 x	y	z	E2 x	y	z	Dia.	Segs
1	16.6	-----	16.6-8=8.6/2=4.3	1	-8.3	0	20	-4	0	20	.1875	2
			4.3+/-4.0=+/-8.3	2	-4	0	20	4	0	20	.25	8
				3	4	0	20	8.3	0	20	.1875	2

Portable dipole, disassembles and stores inside 2 5-foot PVC tubes used as mast.

Fig 1—Example notebook page for a 10-meter dipole.

Band	Antenna Type	# El.	Height	Ground	Date	Run ?
10 M	Yagi (Ant. Bk. 15th Ed p. 11-11)	3	30	R	01 / 12 / 1991	x

L. B. Cebik, W4RNL

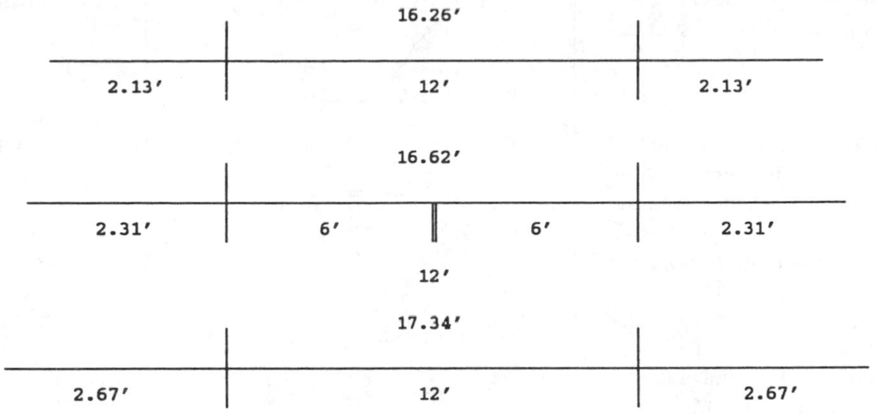

16.26′
2.13′ 12′ 2.13′

16.62′
2.31′ 6′ 6′ 2.31′

12′

17.34′
2.67′ 12′ 2.67′

El. #	Lgth feet	Sp to DrEl	Calculations	Wire #	E1 x	y	z	E2 x	y	z	Dia.	Segs
1	16.62	---	16.62-12=4.62/2=2.31	1	-8.31	0	30	-6	0	30	.875	2
			2.31+/-6=+/-8.31	2	-6	0	30	6	0	30	1.0	8
				3	6	0	30	8.31	0	30	.875	2
2	17.34	5.23	17.34-12=5.34/2=2.67	4	-8.67	-5.23	30	-6	-5.23	30	.875	2
			2.67+/-6=+/-8.67	5	-6	-5.23	30	6	-5.23	30	1.0	8
				6	6	-5.23	30	8.67	-5.23	30	.875	2
3	16.26	3.49	16.26-12=4.26/2=2.13	7	-8.13	3.49	30	-6	3.49	30	.875	2
			2.13+/-6=+/-8.13	8	-6	3.49	30	6	3.49	30	1.0	8
				9	6	3.49	30	8.13	3.49	30	.875	2

Antenna Book design used for program test purposes.

Fig 2—Example notebook page for a 3-element Yagi.

Too, you will be using fractionally dimensioned tubing (for example, 7/8 inch diameter) that you need to enter as a decimal (0.875). And, of course, the calculator serves as a check on addition and subtraction.

Item 5

Multielement antennas remind us to add two more listings. For each element, list the y and z coordinates, the spacing from the driven element and the height above ground. For all single-wire antennas, such as the dipole, set y = 0. Yagis will have other elements separated from the driven element. Each element will be the same distance above ground. Remember to set the source in the center of the driven element for each of the antennas noted here. In order to ensure that the program places your source at the exact center, use an even number of wire segments for the driven element. Otherwise, your source may be offset.

Use the height that you actually anticipate the antenna will be. If you model quads, of course the vertical sections will have changing z coordinates. Once you get used to setting up a few dipoles and Yagis, the changes for vertical sections will come naturally. Start by dividing the vertical section symmetrically and add or subtract from the height of the boom or hub, which you can determine from the real or anticipated tower height.

This completes the notebook page, except for any reminders you may wish to add. I generally enter a note that tells why I modeled the antenna. That refreshes my memory weeks later when I run across the page and wonder why I ever spent computer time on that crazy design.

Building a Baseline of Antenna Information

Let's begin this part of our work with a little scenario. You model a 2-element X beam, using some guesswork and intuition for the dimensions. The program produces the patterns shown in Fig 3. Now, what have we learned? Not much, perhaps, beyond the fact that this is probably not an antenna we'd want to actually build. But you can learn much more by practicing with basic antennas over the soil and land around your own QTH.

This is the reason why, in the program setup, I suggested following the instruction manual and selecting ground conditions that most closely correspond to your own land. I chose average soil for all the examples here because everything indicates that my hilltop QTH on what once was Tennessee farmland is just that: average.

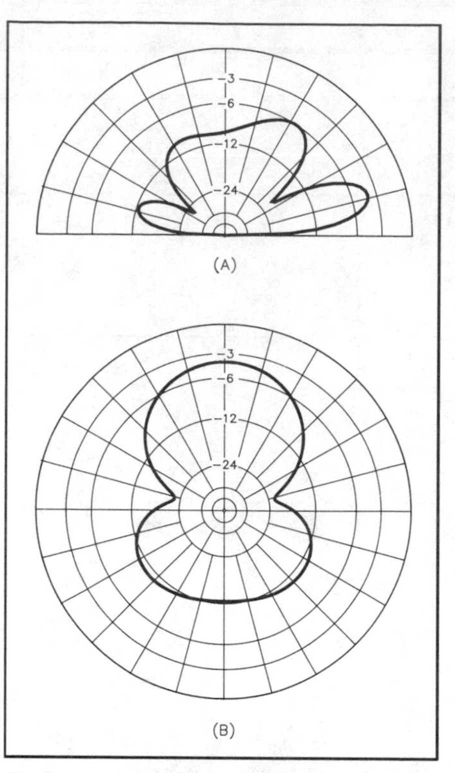

Fig 3—Elevation and azimuth patterns for a 10-meter X beam design, 30 feet high. The maximum gain is calculated to be 9.1 dBi. The azimuth pattern is taken at an elevation angle of 16°. **All patterns in this article were calculated with ELNEC 2.20 for "average" earth having a conductivity of 5 millisiemens per meter and a dielectric constant of 13. In all pattern plots, the 0-db reference (outer ring) is 13 dBi, and all azimuth patterns are taken at the elevation angle of maximum forward gain.**

Once you select and model your ground conditions, stick with them for all models, or your designs may not be directly comparable. If you discover that your soil, rock, or sand differs from your original estimates, then revise and rerun all the antennas that make up your baseline of information.

The next step is to model some basic antennas. Which antennas? You may wish to start with a dipole, a 2-element Yagi, and a 3-element Yagi, as these are all well-established antenna designs with fairly well developed characteristics. Yagi designs are available in many handbooks and ham magazines.

None of these antennas requires more than a single source, and they introduce no loads. Save those complexities until you have mastered the program basics (and then review the manual on how to use them). Moreover, all of these antennas conform to the general idea of avoiding program limitations that may produce misleading results. There are no acute angles for which special techniques are required. Likewise, none of these designs use close-spaced wires, as with folded dipoles. They, too, require special techniques to give accurate results. And none of these basic designs require three or more wires to join at one point.[3] When you are fully comfortable modeling basic designs, then you can add other techniques to your repertoire, one at a time.

Which frequency or frequencies should you use? Select the band in which you are most interested. Later you can expand your baseline to include other bands, especially if you decide to develop multiband antennas. The examples here use 28.2 MHz.

What antenna height or heights should you model? Use realistic heights relative to your situation. For convenience, I modeled the example antennas at 20 and 30 feet, since the lower height is where my temporary dipole is, and the higher elevation is about the size tower I plan to erect. Although two levels are sufficient to illustrate why the baseline data is important, expanding the baseline to 40, 50, and 60-foot heights is not unrealistic.

Heights of at least 20 feet work well for modeling at 10-meter frequencies. However, at lower frequencies (for example, 3.5 MHz), 20 feet would place these horizontal antennas well below the 0.2-λ minimum height for good modeling.[4] Table 1 provides a quick reference to the 0.2-λ height for the HF bands. Below these heights, results are likely to show incorrect

Table 1
0.2 Wavelength at the Lower End of Low-Frequency Amateur Bands

Frequency	*0.2 Wavelength*
1.8 MHz	109 feet
3.5	56
7.0	28
10.1	19.5
14.0	14
18.068	10.9
21.0	9.4*
24.89	7.9
28.0	7

*Heights below this point are not generally useful for modeling, as yard clutter may affect patterns more than program inaccuracies. Mobile, portable, and experimental antenna designs may be exceptions.

impedances and excessive gain. Always keep in mind that MININEC-based programs calculate impedances over a perfect ground, whether you select a perfect or real ground. The results are usable for real ground above the 0.2-λ limit, but may not be usable for less height. Under some circumstances, you may have to erect an antenna below the modeling limit; the antenna may work, but the program may not accurately model its operation.

The three chosen antennas at two elevations each require only six runs of the program. Using constant-diameter elements for the baseline simplifies the design work and lets the computer race through the calculations. More complex antenna construction, of course, gives you more element sections and inevitably longer calculation times.

My early experiences have sold me on the idea of beginning with an elevation pattern. (Elevation is simply the vertical radiation pattern, and azimuth is the horizontal radiation pattern.) Since the x coordinate is in line with the antenna elements, the elevation pattern requires a 90° orientation angle to catch the main lobe. The elevation pattern then tells what angle to use for the azimuth pattern to catch the perimeter of maximum radiation. You can check other azimuth patterns at other angles, but for initial information, a basic elevation and azimuth pattern combination will be instructive. For the purposes of comparison, all the antennas use the same value for the scale on which they are plotted. This permits the omission of detailed analysis information, although you may want that information on the plots you take for your own use.

Figs 4 and 5 illustrate the 10-meter dipole performance at 20- and 30-foot heights. The difference is revealing. The 30-foot-high dipole channels significant energy into high radiation angles, while the 20-foot high dipole appears to concentrate more energy at lower radiation angles. The patterns may come as a surprise if the only patterns you have viewed are free space types. Moreover, the patterns seem to supply an answer to my wondering about the excellent reports from Europe on my 20-foot high 10-meter dipole (without nearby trees). Figs 4B and 5B provide the azimuth patterns of the dipole. They show definite reductions off the ends of the antenna, but not the classic pinch-waisted "figure 8" of free-space patterns. Note that these patterns apply to 10 meters and not necessarily to other bands, because patterns will vary with (among other factors) the percentage of a wavelength the antenna is placed above ground.[5]

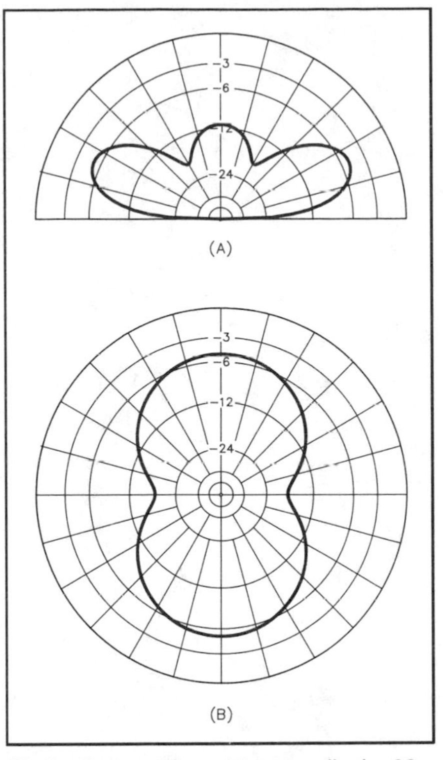

Fig 4—Patterns for a 10-meter dipole, 20 feet high. Azimuth pattern at 24° elevation; maximum gain 8.1 dBi.

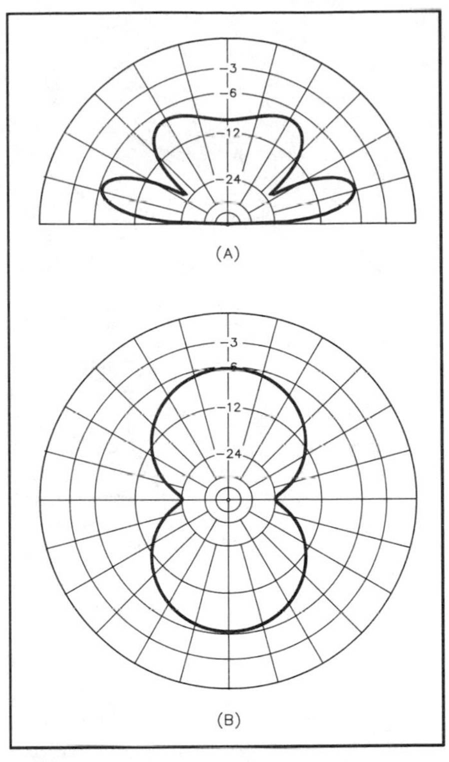

Fig 5—Patterns for a 10-meter dipole, 30 feet high. Azimuth pattern at 16° elevation; maximum gain 6.9 dBi.

It is tempting to compare the 30-foot-high dipole patterns of Fig 5 with Fig 3, the intuitive X beam at 30 feet, and notice the variation of the elevation pattern from the dipole. However, before we can evaluate whether or not the variation—which gives some gain and front-to-back ratio—is good, bad, or indifferent, let's look at the Yagis.

Figs 6 through 9 present patterns for the Yagis at 20 and 30 feet. The 2-element Yagi comes from a design by Bill Orr, W6SAI,[6] while the 3-element Yagi is adapted from *The ARRL Antenna Book*.[7] By comparing the various patterns at equivalent heights, we can see the evolution of the dipole pattern into something with gain and front-to-back ratio. Had we skipped the 2-element design, we might have missed the continuity of pattern development, which shows with special clarity at 30 feet.

The patterns reveal a number of other factors of relevance. First, the Yagis provide significant, but not overwhelming gain relative to real dipoles. (Note: ELNEC provides gain in dBi, gain over an isotropic source. Our interest is in comparing gains of real antennas over real property. To do that, we need only use the *difference* in gain figures to see if we are making significant improvements, and how much.) Only when we combine the gain with the front-to-back ratio do we find the real merits of a beam. The 3-element Yagi shows the most significant increase in front-to-back ratio among the antennas used here for baseline data. A complete analysis of our individual antenna situations would require more standard models, but we have enough here to begin generating expectations. Rational expectations are what good baseline data are designed to give us.

For example, any beam we might propose to build at selected heights should show patterns that compete with the Yagis. We can now see that the intuitive X beam at 30 feet does not do the job. Its pattern is little more than a barely deflected dipole pattern. It does not match even the 2-element Yagi for gain and front-to-back ratio. That does not mean there are not good X beam designs; rather, I came up with a bad beam antenna design and used it for this example. What told me this is a bad design was not a textbook or the program manual, but the baseline information I collected from the antenna modeling program.

The Next Step: Your Own Antennas

Your antenna situation differs from mine, which means that many of the pat-

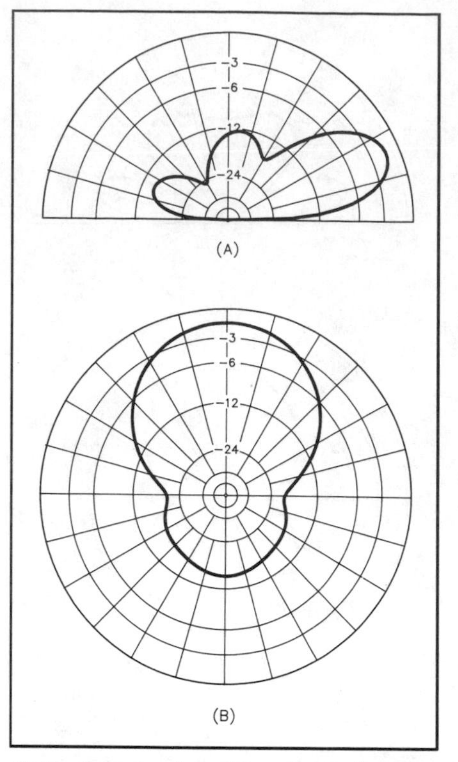

Fig 6—Patterns for a 10-meter 2-element Yagi, 20 feet high. Azimuth pattern at 23° elevation; maximum gain 11.6 dBi.

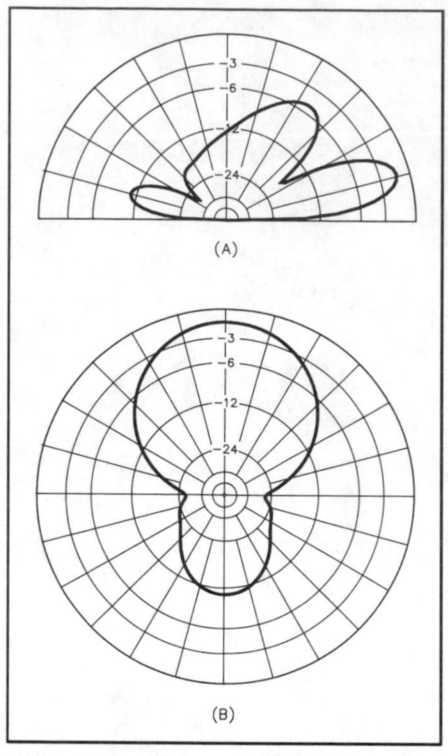

Fig 7—Patterns for a 10-meter 2-element Yagi, 30 feet high. Azimuth pattern at 16° elevation; maximum gain 11.8 dBi.

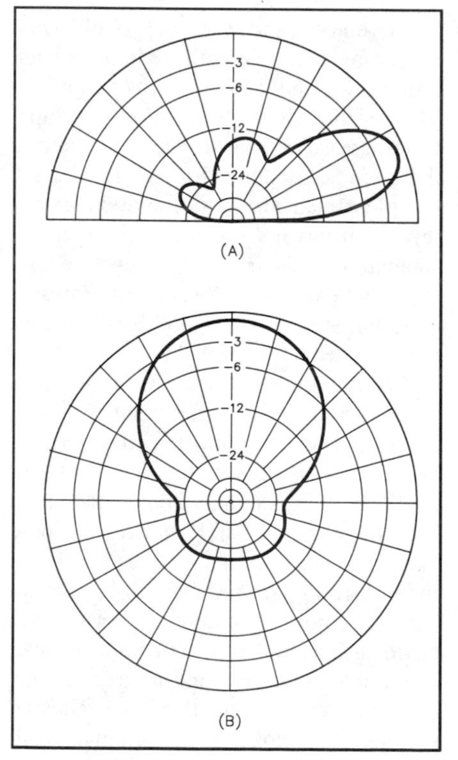

Fig 8—Patterns for a 10-meter 3-element Yagi, 20 feet high. Azimuth pattern taken at 23° elevation; maximum gain 12.3 dBi.

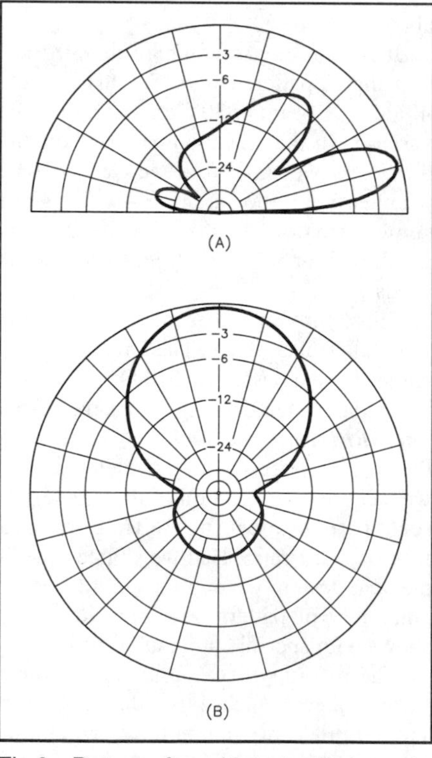

Fig 9—Patterns for a 10-meter 3-element Yagi, 30 feet high. Azimuth pattern taken at 16° elevation; maximum gain 12.6 dBi.

terns shown here may be irrelevant to you. However, they do illustrate the techniques needed by beginners to get the most out of a computer modeling program. The program can be like a textbook written for one person: you. From it, you can put theory into practice without undergoing the expense and time required to build every antenna design you encounter. You can also develop enough models to help you decide where to place your antenna dollar most wisely. Your antenna dollar includes not only the antenna itself, but the mast and tower as well. So even if you practice using the program by modeling the antennas shown here, the next step is to develop a set of models specific to your own needs. A good place to begin is with the antennas you are actually using (unless they contain traps or other complexities that may require advanced techniques to model). The assembly manuals or the calculations you performed to roll your own probably provide almost all the data you need.

Whether you choose your existing antennas or a set of proposed antennas to evaluate with the antenna modeling program, there are a few additional suggestions that may make your effort more profitable. For any antenna, be certain to evaluate it across the entire frequency band of interest. What the band of interest is depends upon your operating interests. If you use the entire band, from the lowest CW segment to the highest phone segment, then you will want antennas that are broadbanded. Their patterns may not show the most gain among your models, but they will show good gain and proper takeoff angles (and, for beams, reasonable front-to-back ratio) over the entire band. If your interests are confined to a narrow portion of any given ham band, then you should check every design at selected spots within that spectrum slice. When making comparisons between designs, use the same set of frequencies for each run.

At this point, the "Source" information becomes important. You will have to be able to match a real antenna to the feed line over the entire band of interest. A high-performance design that will not accept power is not a good antenna for you. Of course, there are many feed-point matching schemes, but choosing one goes outside the program. As you can see, an antenna modeling program is one important part—but not the only part—of an antenna design and construction project or program.

There are many other facets of using antenna modeling programs like ELNEC. The instruction manual is a good guide to them. Working with vertical antennas and ground planes or radial systems, for example, is a topic unto itself with many attached cautions.[8] Accurately modeling real ground requires reference to the program instructions and an understanding of the effects of ground upon antennas.[9] The information presented here is intended to get the beginning antenna modeler started in using the program to learn about antennas, even if he or she never builds one from scratch. Besides all these practical benefits, the programs are fun to use—especially on a rainy day when the bands are closed.

Notes

[1] *MININEC3* by Naval Ocean Systems Center, San Diego, is available from National Technical Information Services, Springfield, VA 22161, order no. ADA181681. This is public domain software, but a generous fee is charged for the diskette and documentation. *MN* by Brian Beezley, K6STI, is available commercially from Brian at 3532 Linda Vista, San Marcos, CA 92069; tel 619-599-4962. *ELNEC* by Roy Lewallen, W7EL, is available commercially from Roy at PO Box 6658, Beaverton, OR 97007. (*MN* and *ELNEC* are enhanced versions of *MININEC3*.) The ARRL in no way warrants these offers.

[2] For a "must-read" description of *MININEC* program limitations, see R. Lewallen, "*MININEC*: The Other Edge of the Sword," later in this chapter.

[3] See the reference in Note 2.

[4] See the reference in Note 2.

[5] [*Editor's note:* Always remember that the gain of an antenna is a function not only of its design, but also its height and the electrical characteristics of the earth. Comparing gain figures of antennas at different heights above ground can be misleading, as Figs 4 and 5 clearly show: The maximum gains differ by 1.2 dB, *for the same antenna*! Of course an antenna cannot have gain over itself. Similarly, front-to-back ratios may change with height at the wave angle of maximum forward response, as Figs 8 and 9 indicate. (The difference for these two heights is 2.1 dB.) As you gain experience with the modeling program, you may want to eliminate any height effects by initially modeling the antennas to be compared in free space (ground-mounted vertical systems excepted). *Then* model your optimized designs over ground to observe their performance in a real-life environment.]

[6] B. Orr, "A Compact 2-Element Yagi for 10 Meters," *CQ*, Dec 1990, pp 83-84.

[7] G. Hall, Ed, *The ARRL Antenna Book*, 15th or 16th eds (Newington: ARRL, 1988 or 1991), design charts on p 11-11.

[8] See the reference in Note 2.

[9] See the reference of note 7, Chapters 3 and 8.

from March 1984 *QST* (Technical Correspondence)

VERTICAL ANTENNAS WORK!

☐ I'm a vertical antenna buff and have been experimenting with them since 1971. My small book (based on the experiments) has sold thousands of copies, and letters from readers prove that, with a common-sense approach, vertical antennas are *not* poor radiators in all directions.

Vertical-antenna enthusiasts should pay close attention to DeMaw's statements in "Building and Using 30-Meter Antennas" (Oct. 1983 *QST*), on p. 29. I want to emphasize that, as operating frequency approaches the upper end of the HF spectrum, antenna height above ground is the difference between cooking earthworms and making contacts. Many letters that substantiate these findings have come to me from operators who have learned to appreciate vertical antennas.

I will "go out on a limb" by saying: Any vertical antenna will perform better (on the 15- and 10-meter bands) mounted 20 ft above ground level with four radials, than the same antenna mounted at ground level with 100 radials. On the 20-meter band, height will most certainly help, but it is not mandatory. On the 40-meter band, there is no difference. Regarding the 75-meter band, I cannot prove that a 65-ft vertical, ground mounted with a tremendous number of radials, will not outperform a top-loaded vertical at 20 ft. If I had the room, I'd like to test it, but on a city lot — forget it!

Antenna height is important on 15 and 10 meters, not only for overcoming obstacles, but also because of the Brewster effect (see "DX and the Brewster Angle," C. L. Hutchinson, Technical Correspondence, May 1983 *QST*), which tends to cancel low-angle radiation at these frequencies. I explained this to a CBer who wrote to me from Kansas. After testing antennas on the ground and at a height of 20 ft, he also found the elevated version superior. As a by-product, he became so fascinated that he soon passed his Novice examination. Chalk one up for our side! — *Charles Schwartzbard, AF2Y, Clifton, New Jersey*

from June 1987 *QST* (Hints and Kinks)

WEATHER PROTECTION FOR VERTICAL-ANTENNA FEED POINTS

☐ I have found that in very wet and rainy climates, where the rain is usually wind driven, a weather boot sealed with silicone caulk does not work well. Coax Seal™ and especially electrician's tape do not adhere well to silicone caulk. The driving rain can be blown or wick into the bottom of the boot, then down the cable braid. Here, however, is an excellent procedure for weatherproofing antenna connections (refer to Fig 4):

- Start with clean, dry connectors (as little skin oil as possible).
- Wrap the connection with several layers of electrical tape.
- Apply Coax Seal to the entire connection and down along the coaxial cable.
- Use two heavy-duty wax-coated paper cups, one inside the other, to make a "tent" around the antenna feed point and coax connector. (The cups deflect rain, so that water doesn't sit on top of the connector joint.)
- Lay a sheet of clear plastic wrap over the cups, push it up inside the cups and then pull it down over the sealed connections.
- Press the plastic wrap into the Coax Seal. (Coax Seal and clear-plastic food wrap stick together, thus forming an air-tight joint.)

Wind, freezing rain and other severe weather cannot damage the joint. The great bond that Coax Seal makes with plastic wrap doesn't allow any moist air or wind-driven rain to enter the joint along the cable surfaces. I've had no leakage problems since using this system.—*James Fox, N7ENI, ARRL Technical Coordinator, Portland, Oregon*

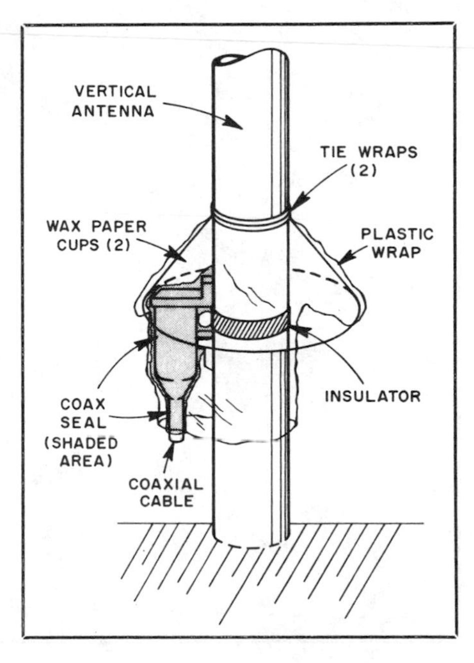

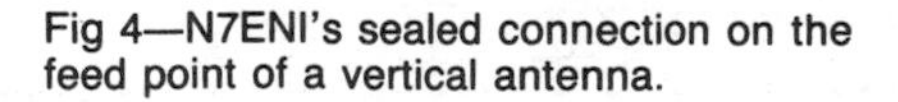
Fig 4—N7ENI's sealed connection on the feed point of a vertical antenna.

from February 1991 *QST*

MININEC: The Other Edge of The Sword

MININEC antenna-modeling software is powerful and popular. But you need to know about its limitations to use it effectively. Here's the lowdown.

By Roy Lewallen, W7EL
5470 SW 152 Ave
Beaverton, OR 97007

Since the dawn of Amateur Radio, *predicting* antenna performance has been justifiably regarded as nearly impossible. No wonder: The only available tools for doing so have been analyses of textbook antennas (that bear a resemblance to our backyard creations to the same extent that a horse resembles a camel), testimonials, folklore and an awesomely lavish dose of "horse puckey."

Now, however, we have been armed with a sharp and powerful sword against decades of antenna-design darkness. But that sword is double-edged and some of us are getting pretty bloody from self-inflicted wounds as we blaze new trails in antenna design.

Our sword is, of course, the powerful antenna-modeling program *MININEC.* One of its edges is its ability to help us answer questions about antennas; its other edge is its limitations which, should we fail to recognize and carefully avoid them, can lead us to conclusions that are embarrassingly and profoundly wrong. For example, from a letter I recently received: "My personal favorite is the 45 dB gain I get [with a dipole] at 0.110 feet [high, over poor ground]. Boy, am I gonna be a big shot on 75 meters now!"

The only error the writer made was not being aware of one of *MININEC*'s basic limitations (discussed later). Tongue firmly in cheek, he had recognized that the answer was ridiculous, but sometimes we're not so lucky and the errors are tougher to spot.

Your ability to avoid the sword's other edge will greatly improve if you take time to gain a basic understanding of what *MININEC* is and how it works.

MININEC was written in BASIC for IBM®-compatible personal computers by J. C. Logan, N6BRF, and J. W. Rockway of the Naval Ocean Systems Center in San Diego. Both the source code and compiled program are available as public-domain software.[1,2] In addition, several commercial programs that use *MININEC* calculation code and additional features have appeared.[3-5] The limitations I'll describe

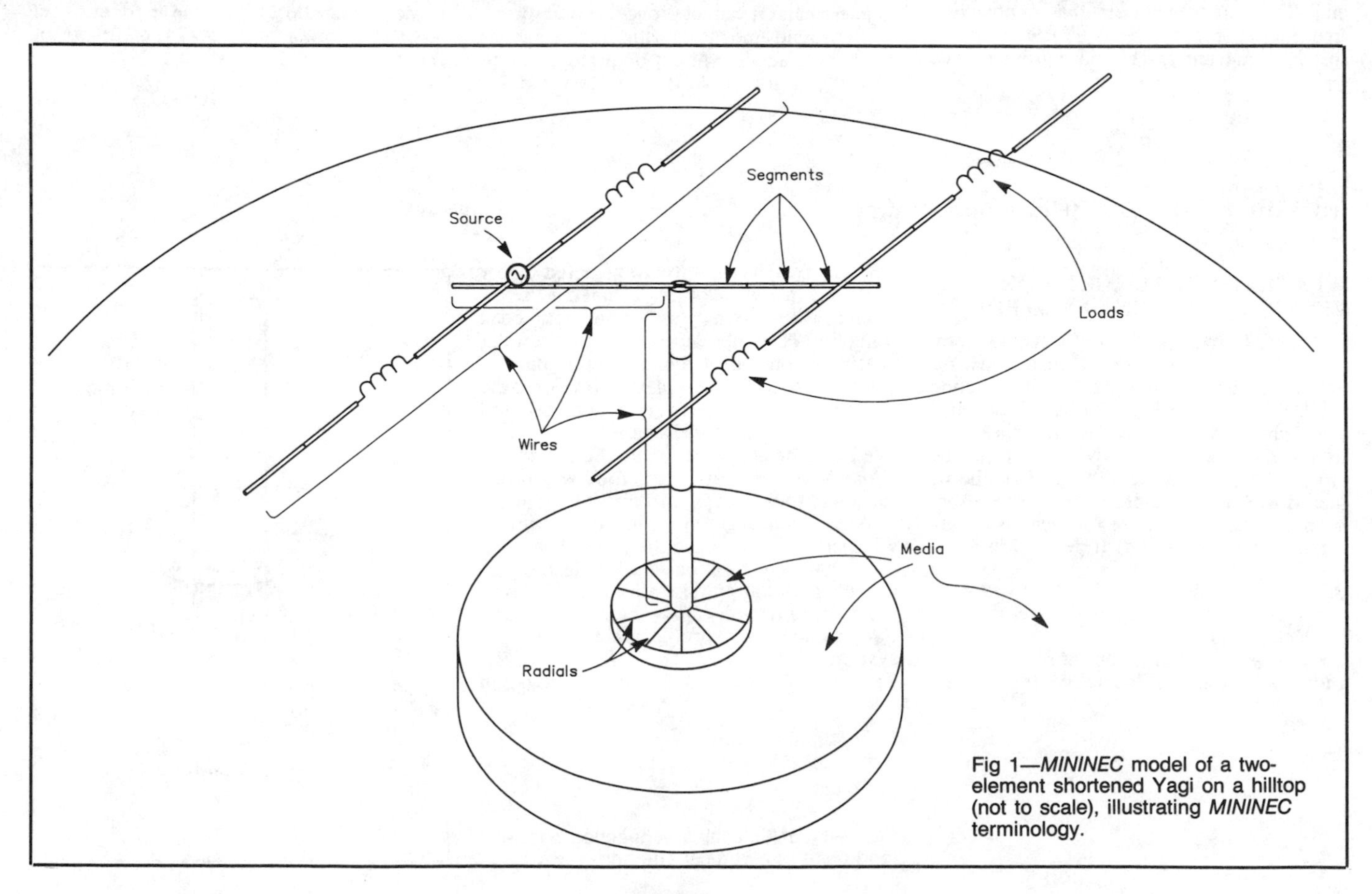

Fig 1—*MININEC* model of a two-element shortened Yagi on a hilltop (not to scale), illustrating *MININEC* terminology.

are, in general, shared by these and other derivative programs. Some variants work around some of the program's limitations, but some also add constraints of their own. Before you use any modeling program, thoroughly *read the documentation* and carefully observe the program's limits.[6] The most important thing you can do is to ask yourself: Does the result *make sense*?

How *MININEC* Works

MININEC is an extremely versatile and powerful program that permits you to "build" an antenna of straight conductors (called *wires*—you choose the diameter), put voltage *sources* and lumped impedances (*loads*) wherever you choose, place the structure over a realistic ground (if desired), and observe the input impedance, current distribution, and near and far fields at any azimuth or elevation angle. (See Fig 1.) Active (driven) and passive (parasitic) structures can be modeled. With some skill and understanding, you can accurately model anything from rhombics to rain gutters and towers to tribanders.

Let's take a closer look at *MININEC*'s operation. You enter the antenna description by specifying the diameters and end points of the wires and the number of *segments* into which they're to be divided for calculation (more about this later). End points are defined in an XYZ coordinate system. A free-space or ground-plane environment can be specified. If you choose a ground plane, it can be perfect or made of one or more sections (*media*) having finite depth, conductivity and permittivity and, if desired, radial wires. Sources and loads can be placed in series with any of the wires. (See Fig 1 for an example.) After entering the antenna description, you select one of several analysis options.

MININEC uses a procedure known as the *method of moments.*[7] In *MININEC,* each wire is divided into a group of equal-length segments for calculation. A uniform current is assumed to flow in a region extending to both sides of each segment junction (see Fig 2). These regions of uniform current, centered about the segment junctions, are called *pulses.* In any analysis, the program first calculates the self-impedance at each pulse and the mutual impedance between each pulse and all the others. If a ground plane has been specified, the impedances to and from the "image" antenna created by the ground plane are also calculated. This operation consumes the majority of the total computation time, reported by the program as *fill matrix.* The result is an internally stored matrix of impedance values. The program then solves an Ohm's Law equation using the values in the impedance matrix, a source-voltages matrix, and a matrix of the unknown pulse currents, reporting *factor matrix.* After this step, the impedances seen by the sources, as well as the currents at each pulse, are available. If near- or far-field analysis is requested, the contribution

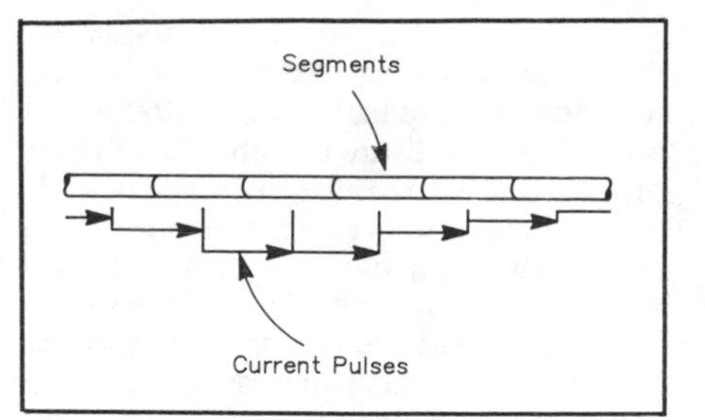

Fig 2—Illustration of the relationship between segments and pulses.

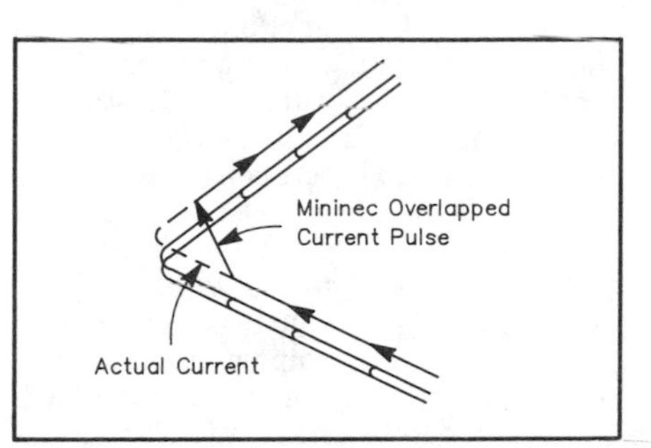

Fig 3—Pulse overlap at wire junctions can cause problems if it's not accounted for.

of each current pulse to the total field is calculated. If a ground plane has been specified, direct and reflected rays are summed to obtain the total field strength at each point of the near or far field.

The Limitations

MININEC's authors did an amazing and commendable job of reducing some very complex mathematical operations to a level that a PC can handle in a reasonable amount of time. But to do so, they had to make some compromises. Most of the program's limitations are due to these consciously chosen compromises.

Wires

In *MININEC,* every antenna must be described using only *straight wires* as the basic model building block. With some ingenuity, though, a wide variety of structures, including towers, top hats, rotators, rain gutters and even garages, can be adequately modeled. But overlapping wires *aren't* automatically connected by the program. For example, four wires are required to model an **X**-shaped structure if the conductors are connected at the center of the **X**. No limit is imposed on the minimum wire radius, and the program will produce accurate results with wire radii as large as 0.01 λ.

Number of Segments

It's up to you to decide how many segments to break each wire into for analysis purposes. To make an appropriate choice you have to have some knowledge of the trade-offs involved. Because the results become more accurate as the number of segments is increased, *MININEC* users naturally tend to use a large number of segments. Two factors suggest caution here. First, the size of the complex-impedance matrix calculated by the program goes up as the *square* of the number of pulses. (The number of pulses is approximately equal to the number of segments.) Therefore, *MININEC* and all its derivatives have some limit on the allowable number of pulses. Second, analysis time increases approximately as the square of the number of segments.

So, just how many segments are required to "do it right"? There's no exact answer, because the analysis accuracy nearly always improves with more segments. A straightforward (but time-consuming) way to determine if you've used enough segments is to increase the number of segments, rerun the analysis and see how much the results change. Some rules of thumb work well and can be used as a starting point if particularly good accuracy is required. As I'll describe, you need to take special care at wire junctions, especially where wires are connected at an acute angle.

Straight Wires

If you want to look at the pattern of an antenna with straight elements (like a Yagi), eight to ten segments per half wavelength are adequate. The pattern won't change much as you increase the number, although the program may give more accurate null depths with more segments. If you require *really* accurate feed-point impedances, use more segments.

Connected Wires—General

It's easiest to understand some of the problems of connecting wires if you have an understanding of what *MININEC* does at wire junctions. An unconnected wire is left with a zero-amplitude half-pulse at its end. However, the end pulses of later-defined connecting wires have nonzero current amplitudes. The half-pulse that extends beyond one of these wires is overlapped onto the lowest-numbered connecting wire (Fig 3). *This half-pulse of current takes on the segment length and wire diameter of the lower-numbered wire.*

When the program does its calculations, it considers only the pulse center or end points. When the straight-line path between the pulse ends becomes substantially different from the actual current path, errors result. This occurs wherever wires are connected at a nonzero angle. Accuracy also suffers when wires having greatly different segment lengths are connected. John Belrose, VE2CV, has observed[8] that the best results are obtained with square loops when the segment lengths are the same on all legs. He also observes that, as a rule, segment lengths on connected wires should differ by no more than a factor of two. Both rules are reasonable considering the way *MININEC* handles connections, and both rules have experimentally been proven sound.

Table 1

Feed-Point Impedances Reported by *MININEC*†

Straight Dipole

Segments	*Impedance (ohms)*
10	74.073 +*j* 20.292
20	75.870 +*j* 21.877
30	76.573 +*j* 23.218
40	76.972 +*j* 24.053
50	77.222 +*j* 24.517

Bent Dipole

Segments	*Impedance (ohms)*
10	11.509 −*j* 76.933
20	11.751 −*j* 53.812
30	11.819 −*j* 46.934
40	11.848 −*j* 43.783
50	11.861 −*j* 41.988
14††	11.312 −*j* 43.119

†Impedances for a straight 0.5-λ dipole and dipole bent horizontally at its center at a 45° included angle, with various numbers of segments. Both antennas have a wire radius of 0.001 λ and are placed 0.5 λ above perfectly conducting ground.

††Tapered segment length. See text.

Wires Connected at Right Angles

Wires connected at a nonzero angle require more segments than unconnected wires or those connected at a 0° angle. Eight to ten segments per half wavelength are required for reasonable results if the connection angle is 90° or less and the segment lengths of both wires are equal. Far-field accuracy of a one-wavelength-circumference square loop is reasonably good with four segments per leg, although once again the impedance accuracy improves with more segments.

Wires Connected at Acute Angles

This is where *MININEC* becomes tricky. Accuracy can rapidly degrade as wire-connection angles decrease, although here again the impedance loses more accuracy than the far-field pattern. An example is shown in Table 1. The *MININEC*-calculated impedance of a dipole is reasonably accurate when the antenna is divided into only ten segments. When the same dipole is bent at a 45° included angle, *more than 30 segments are required for similar accuracy.* In both cases, however, ten segments produce far-field patterns that are virtually indistinguishable from those produced using more segments. The only way I know of to evaluate these cases is to change the number of segments and see what happens. Described next, however, is a technique that you can use to reduce the number of segments required for wires connected at sharp angles.

A Technique for Improving Accuracy

MININEC's accuracy can be markedly improved at wire junctions with only a slight increase in the total number of segments. This is done by tapering the segment length, making it short in the vicinity of the wire junction and increasing it at greater distances. Typically, only a few extra wires are required. This technique is illustrated in Fig 4 for the bent dipole of Table 1. Wires 1, 2 and 3 and their counterparts on the other half of the dipole have only one segment each. The remainder of the half-dipole is one four-segment wire. These segments are just slightly longer than when the dipole was made up of 10 segments total. The net result, shown in the last row of Table 1, is that the impedance for this 14-segment model is similar to the 40-uniform-segment model.

Close-Spaced Wires

MININEC documentation includes analysis of parallel wires at various spacings and finds the program to be well-behaved even when wires are very close together. Nonetheless, it cautions, "Whenever a model has close spacing, however, it is advisable to examine the results very closely to ensure proper behavior." Some time ago I analyzed a typical open-wire transmission line and found it necessary to make the segments no longer than three times the wire spacing. With longer segments, dramatic impedance errors resulted. More recent experiments have indicated that the problem is caused not by the close spacing, but by the connection at the ends of the two wires. The wire connecting the end has a maximum possible segment length equal to the wire spacing. The rule of having no more than a 2:1 segment-length ratio (see *Connected Wires—General*) on connected wires is violated unless the main wire-segment lengths are no more than twice the wire spacing. The tapered-segment-length approach outlined above can be successfully applied in this situation.

Additional factors limit your choice of the number of segments to use. Because *MININEC* assumes that current is uniform along a pulse, segment lengths should be short enough that the current in the real antenna doesn't change much in this distance. Therefore, the maximum segment length shouldn't exceed about 0.1 λ. *MININEC* documentation also states that segment length should always be greater than 10^{-4} λ, and greater than 2.5 times the wire radius.

Sources and Loads at Multiwire Junctions

This one can be a real surprise. When you place a source or load at a junction of more than two wires, you have to be very careful, or the source or load won't end up where you thought! Sources and loads can be placed only at pulses (segment junctions), so to understand the problem you need to know how *MININEC* assigns pulse numbers. Here are the rules it uses:

See Fig 5. Pulse numbering begins at end number 1 of wire number 1. A pulse number is assigned to each segment junction on the wire, and at a wire end if the end is connected to ground or an already-defined wire. No pulse numbers are assigned to open wire ends. After pulse numbers are assigned to the first wire, pulses are assigned to wire number 2, again beginning at end 1, and so forth.

Wire 1, shown by itself in Fig 5A, has four segments to which three pulses are assigned. Pulse numbers 1-3 belong to wire 1. In Fig 5B, wire 2 is added. Note the assignment of pulse number 4, which belongs to wire 2 since it didn't exist until wire 2 was defined. When wire 3 is added in Fig 5C, pulse number 8 is assigned *to the same physical junction as pulse number 4,* in accordance with the above rule. Pulse 8 belongs to wire 3. Now suppose the antenna in Fig 5 is a groundplane with two drooping radials, and we want to place a source at the base of wire 1, the main radiating portion. If we specify pulse 4 for the source position, *the source ends up on wire 2,* as shown in Fig 6. If the source is producing I amperes, I amperes flows in wire 2, and the return current of I amperes splits between wires 1 and 3—not the desired result. The same thing happens if pulse 8 (Fig 5C) is specified, except that wire 3 gets the full current and the return current splits between wires 1 and 2. Putting the source at pulse number 1 gets it on wire 1 all right, but 25% of the way up from the junction. *There is no way to place the source on wire 1 at the wire junction as this antenna has been defined.* This is because there's no pulse belonging to wire 1 at the bottom (end 1) of wire 1. *The only way to achieve the desired result is to avoid placing the source in the lowest-*

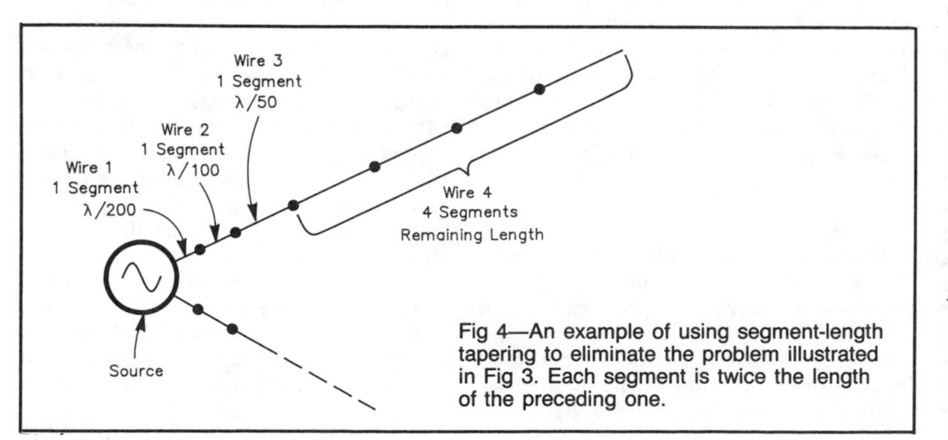

Fig 4—An example of using segment-length tapering to eliminate the problem illustrated in Fig 3. Each segment is twice the length of the preceding one.

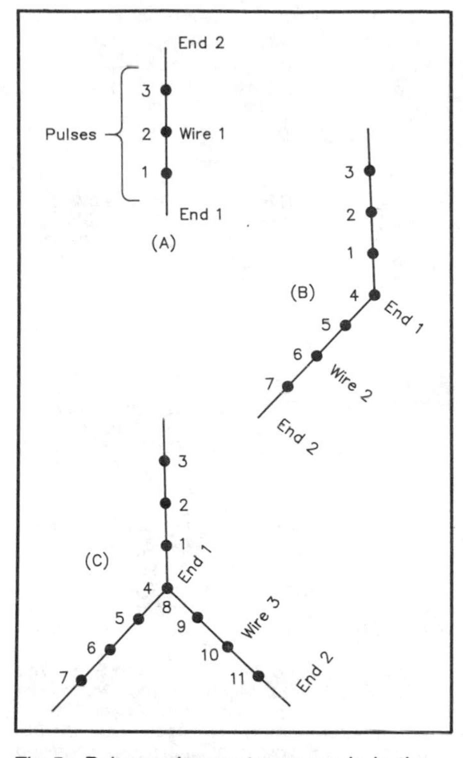

Fig 5—Pulse assignments are made in the order that wires are defined. See text for explanation.

numbered wire (the one defined first) in the group sharing a common junction. To make sure you've put the source where you think you have, always look at the currents and *make sure they make sense.* Load placement behaves the same way, but mistakes can be harder to spot, so be extra careful when placing loads at multiwire junctions.

When only two wires are connected, there's no problem. Regardless of which wire the common pulse belongs to, the entire source or load current flows in both wires.

Ground—General

Probably the most misunderstood limitation of *MININEC* is its ground-modeling capability. Even though the program permits you to define a real ground in considerable detail, this definition is used *only for calculating far-field patterns. MININEC* uses *perfectly conducting ground* when calculating impedances and currents if either a perfect or real ground is specified. Ground has several effects on antennas.[9] Let's look at them one at a time and see how this simplification affects the accuracy of results.

Impedance and Gain

The feed-point impedance of an antenna changes with antenna height. The magnitude of this effect depends on antenna length, diameter, and orientation, and the ground characteristics. The impedance change of a half-wave dipole above ground is well documented.[10] When a dipole is at least 0.2 λ above ground, its impedance is

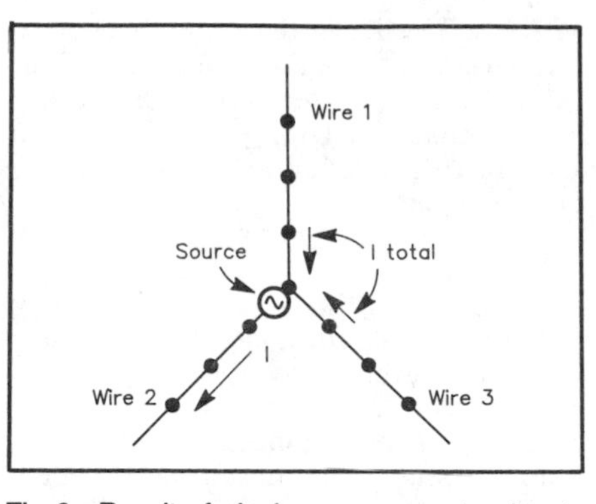

Fig 6—Result of placing a source at pulse 4 on the antenna shown in Fig 5C.

nearly the same whether the ground is real or perfect, so *MININEC* results are adequate. If the antenna is lower, however, *MININEC* results can deviate greatly from the true impedance of an antenna over real ground. Specifically, the resistance reported by *MININEC* will be lower than it really is. This in turn leads to excessively high reported gain, as noted by the correspondent quoted earlier. I don't have any information on longer dipoles (such as the extended double Zepp), but I suspect that these antennas must be somewhat higher than 0.2 λ before *MININEC* gives accurate impedance results. Vertical dipoles have about the same impedance over real ground as over perfect ground, so *MININEC* results are satisfactory for these antennas at any height.

Ground System Losses and Efficiency

Efficiency of ground-mounted low-impedance vertical antennas is frequently the most severe limitation on such an antenna's performance. In such antennas, power is lost due to feed-point return current flowing through lossy ground in the vicinity of the antenna base. Placing radial wires around the base of the antenna raises efficiency by reducing this loss. A common way to determine the loss in such an antenna is to measure the feed-point resistance and compare it with the resistance of a similar element over lossless ground. The loss is simply the difference in the resistances, and usually this is nearly all ground loss.

Because *MININEC* uses a perfect ground for impedance calculations, it always reports the impedance seen over a perfect ground. Therefore, *MININEC* can't be used to determine the impedance or efficiency of antennas fed against ground; you can't use the program to evaluate the effectiveness of radial systems, for instance.

Low-Angle-Radiation Attenuation

Vertically polarized waves, in particular, are attenuated when reflected from lossy ground, leading to the well-known phenomenon of low-angle radiation attenuation.[11] *MININEC* models this correctly as part of the far-field calculation. If radials are specified, they modify the conductivity of the ground on which they are placed. *MININEC* documentation cautions that the radial calculations are accurate only for large numbers of radials.

One additional caution is necessary when specifying ground constants. It's frequently convenient to model antennas at 299.8 MHz, where a wavelength is one meter. If you do this, *you must scale ground conductivity in proportion to the frequency.*[12] For example, an antenna operating at 7 MHz over ground having a conductivity of 0.002 S/m will behave like a size-scaled antenna operating at 299.8 MHz over ground with $0.002 \times (299.8 \div 7) = 0.086$ S/m conductivity. However, even with a ratio this large, neglecting to scale ground conductivity usually won't be apparent in the far-field patterns, except at very low angles.

Multiple Media

Other limitations appear when the ground is broken up into several pieces (*media*). Once again, it's helpful to understand how *MININEC* functions in this regard.

Height of Ground Under the Antenna

It's important to realize that *MININEC* always assumes a ground-plane height of 0 (Z = 0 in *MININEC*'s XYZ coordinate system) when calculating impedances and currents. Also, it regards a wire-end Z coordinate of zero as meaning that the wire is connected to ground (except when the antenna is being modeled in free space). For these reasons, the region of the ground plane immediately under the antenna must have a height (Z coordinate) of zero. If you're modeling an antenna on top of a hill, the top of the hill must have a Z coordinate of zero, with the rest of the hill having negative Z coordinates.

Other Concerns

At each elevation angle, *MININEC* looks for reflection from ground. It begins at the most distant medium and looks for the intersection of the direct and reflected rays. This process is repeated for all other media. *MININEC* uses the innermost reflection point it finds; it makes no attempt to evaluate multiple reflections or those from corners. *MININEC* doesn't look between the antenna and the reflecting point or beyond the reflecting point. Therefore, the program assumes that RF passes through hills and cliff walls with no shielding or reflections. A puzzling simplification is that the program assumes a height of zero for all media during the process of determining the wave-reflection point to be used for far-field calculations, although media height is taken into account during summation of the incident and reflected rays. This can lead to pattern errors with media of differing heights. If you have access to a compiler, you can easily patch *MININEC*'s source code to overcome this deficiency. See the appendix for details.

These ground approximations were purposely made to keep the program length and speed compatible with PCs. We can hope that improved ground-modeling code will become available in the future as PCs continue to increase in speed and power.

Loss

MININEC doesn't automatically account for loss. Therefore, be wary of antennas with low feed-point resistances. The answers might be entirely legitimate, but only if there is no loss in the antenna structure. In the lossy real world, these antennas just won't work. Whenever you see a surprisingly high gain, look at the feed-point resistance and you're likely to find it's very low. Imitate reality by adding loads having a few ohms of resistance at each source (and anywhere else the current is high) and watch what happens to the gain!

Frequency-Related Errors

At least two writers have reported apparent frequency-dependent errors in *MININEC.*[13,14] This was determined by comparing *MININEC* results to those of *NEC*, a much more sophisticated mainframe program. Their observations were that, for certain frequencies and element diameters, the two programs seem to give similar results at slightly different frequencies. The only specific example of this I've seen was provided by Peter Beyer, PA3AEF.[15] It shows *NEC* and *MININEC* analyses of a 10-element 144-MHz Yagi. The *NEC* analysis was done at 144.5 MHz. *MININEC* analysis is closer to the *NEC* results when done at 145 MHz than at 144.5 MHz. I ran some brief experiments to see if there is, indeed, a frequency-sensitive error *within MININEC.* I scaled the same antenna for different frequencies and analyzed them with *MININEC.* No frequency-dependent effects (resonance shift, etc) were found, but the tests were far from exhaustive, and the program's *absolute* accuracy is what's in question. My feeling is that the differences arise because of the much more sophisticated way in which *NEC* deals with currents. I hope we'll see more about this phenomenon in amateur publications. In the meantime, be careful when trying to get high accuracy from *MININEC* analysis of highly directional structures, especially at VHF and UHF.

Bugs

I know of only two actual bugs in *MININEC.* They both deal with Laplace ("S-parameter") loads. One causes an overflow and the other is very obscure and highly unlikely to affect you. If you'd like some further description and fixes for the bugs, contact me.

Summary

All modeling tools, no matter how elaborate, powerful and expensive, have limitations. Absolutely none of these can be used sensibly unless you're constantly conscious of their limitations. *MININEC* is no exception. You must always be alert for answers that don't seem quite right. Are the impedance and gain values *reasonable?* If the antenna is symmetrical, is the pattern symmetrical about the axis you intended to specify? Do the currents change abruptly from one segment to another?[16] Do the results seem too good to be true? *If so, they probably are!*

We owe *MININEC*'s authors a great debt of gratitude for the pioneering work they have done. They've put fast, accurate antenna analysis within the reach of thousands of amateurs. The program they have created is very useful for analyzing a variety of antenna designs. Wielded properly, *MININEC* can be a powerful tool—a weapon against a decades-long void in knowledge about antenna design. This article should help you avoid the other edge of the sword.

APPENDIX

If you have access to BASIC compiler software (eg, Microsoft® *QuickBasic,* Borland *Turbo Basic*), you can patch the *MININEC* source code to improve *MININEC's* handling of multiple media of different heights, then recompile the program.† Of course, the source code could be run directly with a GWBASIC interpreter, but the speed will be so slow as to render the program virtually useless. In the following code segments, the added lines have no line numbers since such are not required by the compilers.

```
702 T3 = -SIN(U4)
IF ABS(R3) < 0.00001 THEN ATU4 =
   100000 ELSE ATU4 = ABS(T3/R3)
703 T1 = R3 * V2
756 FOR J1 = NM to 1 STEP -1
IF B9 > U(J1) + H(J1) * ATU4 THEN 759
758 J2 = J1
```

Note: Delete line 757
[IF B9 > U(J1) THEN 759].

If the program is to be compiled with Microsoft *QuickBasic,* one other change must be made. In *MININEC,* "IS" is used as a variable. Because "IS" is a reserved word in *QuickBasic,* it must be changed. (If you're using a different compiler, check its documentation to see if this change is required.) Change "IS" to "ISX" in the following lines: 1592, 1593, 1596, 1605-1609, and 1612.

†Patched *MININEC* in compiled form is available from the author on an MS-DOS 5¼- or 3½-inch disk for $3 postpaid to the US, Canada, and Mexico. Add $3 airmail postage to other countries.

Notes

[1]*MININEC* is available from National Technical Information Service (NTIS), US Department of Commerce, 5285 Port Royal Rd, Springfield, VA 22161, tel 703-487-4650. Order no. ADA181681 (software and documentation).

[2]A technical reference describing the program is J. C. Logan and J. W. Rockway, *The New* MININEC *(Version 3): A Mini-Numerical Electromagnetic Code,* NOSC TD 938, Naval Ocean Systems Center, San Diego, CA, 1986. It is available as document number ADA181682 from NTIS (see note 1). This is a highly technical manual.

[3]J. Rockway, J. Logan, D. Tam and S. Li, *The* MININEC *System: Microcomputer Analysis of Wire Antennas,* available from Artech House, 685 Canton Street, Norwood, MA 02062. Includes several programs with source code and a comprehensive manual.

[4]*MN* and *MNjr*, by Brian Beezley, K6STI. Available from Brian Beezley, 3532 Linda Vista, San Marcos, CA 92069; tel 619-599-4962.

[5]*ELNEC,* by Roy Lewallen, W7EL. Available from Roy Lewallen, PO Box 6658, Beaverton, OR 97007.

[6]Documentation files for *MN* and *ELNEC* are available on 5.25-inch diskettes from their authors for $5 and $3, respectively. Add $3 for postage to locations outside North America. See notes 4 and 5 for addresses.

[7]A good description of the method of moments is included in J. D. Kraus, *Antennas,* 2nd edition (New York: McGraw-Hill, 1988), pp 359-408.

[8]J. S. Belrose, VE2CV, ARRL Technical Advisor, private correspondence.

[9]An excellent description of these effects appears in G. L. Hall, ed., *The ARRL Antenna Book,* 15th edition (Newington: ARRL, 1988), Chapter 3.

[10]See note 9, p 3-11, Fig 16.

[11]See note 9, pp 3-1 through 3-6 and 3-10.

[12]G. Sinclair, "Theory of Models of Electromagnetic Systems," *Proceedings of the IRE,* Nov 1948, pp 1364-1370.

[13]P. Beyer, "Antenna Simulation Software," *Proceedings of the Third International EME Conference,* Thorn, Netherlands, Sep 9-11, 1988. Thanks to Warren Butler, W2WD, for bringing this to my attention.

[14]R. Cox, "An Update on Computer-Aided Antenna Design," *1990 Central States VHF Conference Proceedings,* published by ARRL. Thanks to *QST* Assistant Technical Editor Rus Healy, NJ2L, for bringing this to my attention.

[15]Peter Beyer, PA3AEF, private correspondence.

[16]Positive current flow is defined as being from end 1 to end 2. Current reversals at wire junctions are normal if wires are connected "head to head," ie, end 1 to end 1 or end 2 to end 2.

Chapter 2

VHF and UHF

An Investigation Of 2-Meter HT Antenna Performance

By P. K. (Ken) Pierpont, KF4OW
204 Cedar Point Crescent
Yorktown, VA 23692

Ed Brummer, W4RTZ
108 Oyster Cove Road
Yorktown, VA 23692

Should you change the antenna on your HT? First take a look at what you really get with different antenna types and sizes. These measurements were taken under realistic conditions.

Introduction

In 1984 one of the authors (Ken, KF4OW) received his first 2-meter HT. He quickly recognized the 1-W unit didn't provide the communication desired despite its compactness, portability and convenience. An improved antenna seemed a logical first step. The first improved antenna was a λ/4 whip. It was constructed from a BNC connector with a piece of piano wire potted in place with epoxy. A thumbtack was soldered on top as a safety measure.

Next, after studying the ads and reading as much information on the subject as could be found, a commercial λ/2 telescoping antenna was purchased. A discussion with coauthor Ed Brummer, W4RTZ, led to the idea of making a few field tests to compare performance. A 5/8-λ mobile whip was borrowed to make an additional comparison with the λ/2 antenna. The first field experiments included attaching the HT to an existing wood post with rubber bands. The post was conveniently located about a mile and a half from the receiving station, and S-meter readings were used as the performance indicators. The results of these tests, performed in February, contained many anomalies and therefore were inconclusive.

In early March a second attempt was made to take an improved set of measurements. For these tests, an adjustable wood mount was built for the HT. This mount would permit rotation through 360° of azimuth angle and inclinations of the HT from antenna vertical to about 45°. Two additional popular telescoping 5/8 λ antennas, intended for HT use, were included in this second series of tests. All the antennas were tested over a range of azimuth angles and tilt angles with respect to the mouth of the operator in a normal position close to the built-in microphone.

A number of things were learned from these first two field experiments. These lessons were then employed in a final set of tests:

1. Do be aware of the difficulties in making antenna performance measurements. *The ARRL Antenna Book*[1] has some words of wisdom on this subject.

2. Be sure of the calibration of any meter used to indicate signal strength, and always take repeated reference measurements.

3. The test range must be clear of potential RF obstructions, such as trees, houses, power lines, etc.

4. Use a mounting system known to be transparent to RF at the frequencies of interest. Some plywoods, glue bonds and solid wood posts, as well as other materials like plastics, may not be truly RF-transparent. All materials proposed for use should be tested.

5. Use only a single power supply and take numerous reference data points to ensure uniform transmission levels. Transmission times should be brief to conserve battery power. The power level will be kept the same if only a fraction of the battery capacity is used. Never change battery packs during a test run.

6. Anything attached to the HT, such as a portable mike or auxiliary power cable will affect the characteristics of the emission patterns and should be avoided.

7. The presence of the operator's body will produce large effects on the emission pattern and must be controlled carefully.

A new mounting system was constructed based on these principles. RF transparent materials were used and an improved test range site was selected for the third series of tests. At the end of March, when the weather improved, a new test procedure was used to obtain what we believe are reproducible results.

Experimental Apparatus

Antenna Configurations Investigated:

The antenna configurations tested consisted of a helical-coil rubber duck supplied by the HT manufacturer, a home brew $\lambda/4$ whip antenna made of piano wire, and three commercially available telescoping antennas. One telescoping antenna was $\lambda/2$ when extended and the other two were $^{5}/_{8}\,\lambda$ when extended. All three telescoping antennas were claimed by their manufacturers to be usable when fully collapsed and therefore were also evaluated in the collapsed configuration.

Eight antennas were measured. They are shown in **Fig 1**. Because the purpose of this investigation was to evaluate performance for communication, no consideration was given to any other advantages claimed by a manufacturer.

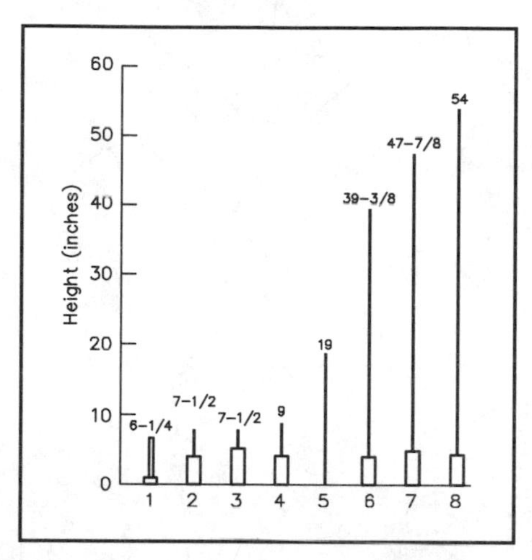

Fig 1—Description of HT antennas tested. All dimensions are measured from the top of the BNC connector. Antenna 1 is the reference rubber duck. Antennas 2, 3 and 4 are collapsed forms of telescoping antennas. Antenna 5 is a 19-inch whip, and antennas 6, 7 and 8 are telescoping antennas.

Transmitting and Receiving Equipment:

The test transmitter was a 2-meter HT capable of putting out both 1 W and 100 mW power levels with its standard battery pack. The rubber duck helical antenna supplied by the manufacturer was used as the reference antenna and is called the baseline configuration in this article. A reliable 2-meter transceiver equipped with a built-in S-meter and a 10-dB attenuator was used for received-signal measurements. Mounted on the roof at a height of about 25 feet was a homemade J-pole. It was matched to the feed line and transceiver with an SWR better than 1.1:1.

Field HT Mount:

The HT field-mount system is shown in the sketch of **Fig 2**. It was constructed of $^{3}/_{4}$-inch and 1-inch PVC pipe and clear $^{1}/_{8}$-inch plastic sheet to be entirely RF-transparent at these frequencies. All materials were tested to ensure their RF performance. The mount itself was a glued box, to which was attached the HT belt clip. A pivot made from $^{1}/_{4}$-inch PVC tubing and a second PVC tubing tilt-angle adjustment pin provides forward tilt from an initial vertical position of $\phi=0°$ to 45° in 15° increments. The upper-post section was 1-inch PVC pipe telescoped onto a $^{3}/_{4}$-inch section reinforced with a wood dowel inserted about a foot into a hole in the earth. Vertical adjustment was provided by a series of holes and an adjustment pin made from $^{1}/_{4}$-inch PVC tubing.

The azimuth angle was changed by rotating the entire mount. Optical alignment for direction was made with a reference compass rose laid out on the ground. Zero degrees was defined as a line connecting the receiving and transmitting stations, as determined by compass reading based on a US Coast and Geodetic Survey chart. The two primary angles of azimuth and tilt are shown in the sketch of **Fig 3**. The operator's position relative to the HT is also shown in Figs 2 and 3. In all cases, the operator head, body, arms, and hands were carefully kept in the same relative positions. Even the location of the hand and fingers affected the pattern. The importance of these factors cannot be overemphasized if reliable results are to be obtained and repeated.

Test Range:

The map of **Fig 4** shows the test range

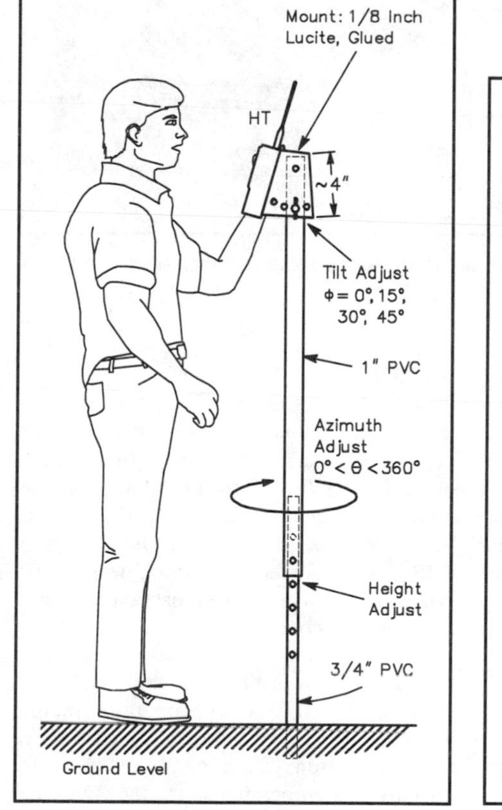

Fig 2—Field mount system.

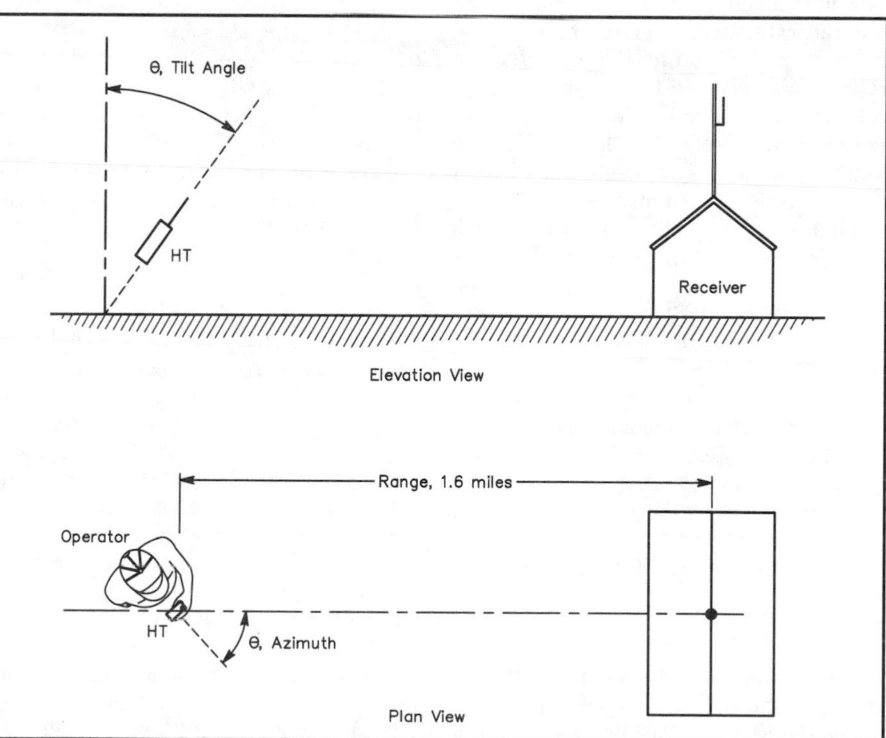

Fig 3—Definitions of azimuth and tilt angles.

used. The receiving antenna was in a clear area. The transmitting location was free of any obstructions, or RF reflectors, for at least 100 yards in any direction and more than 300 yards toward the receiving station. The nearest tree was about 150 yards to the side. The transmitting setup was located near a large pit filled with water to form a small lake, extending about 1/4 mile ahead and on both sides of the operations site.

Test Measurements and Data Reduction

All measurements were made at the receiving station by carefully determining the average value indicated by the S meter during an approximately 10 second unmodulated FM transmission. Since the distance between the receiver and transmitter was approximately 1.6 mi, the transmitter power level was kept at 100 mW. An attenuator was used to keep the received signal in the mid-scale range of the S meter. Each test run consisted of a single antenna configuration at a given angle of tilt. Data were taken for a series of azimuth angles by rotating the transmitter and operator through the 360° range. A repeat zero was always taken, and both initial and final data were taken for identical test conditions to confirm battery power level changes had not affected the results. In several cases, two test runs were made just to obtain typical data-scatter information.

A before- and after-test calibration of the S meter was made using a precision step attenuator. Thus the accuracies of all earlier calibrations were repeatedly confirmed. All data have been reduced by means of the calibration factors to a relative decibel scale using the zero-signal level as the reference value. This relative data is presented on polar coordinate graphs. Constant values (in azimuth) are therefore circles. In **Fig 5** the 5 and 10-dB circles are labeled and circles indicate 2-dB increments. Each configuration is assigned a symbol, and repeated data are represented by symbols with flags.

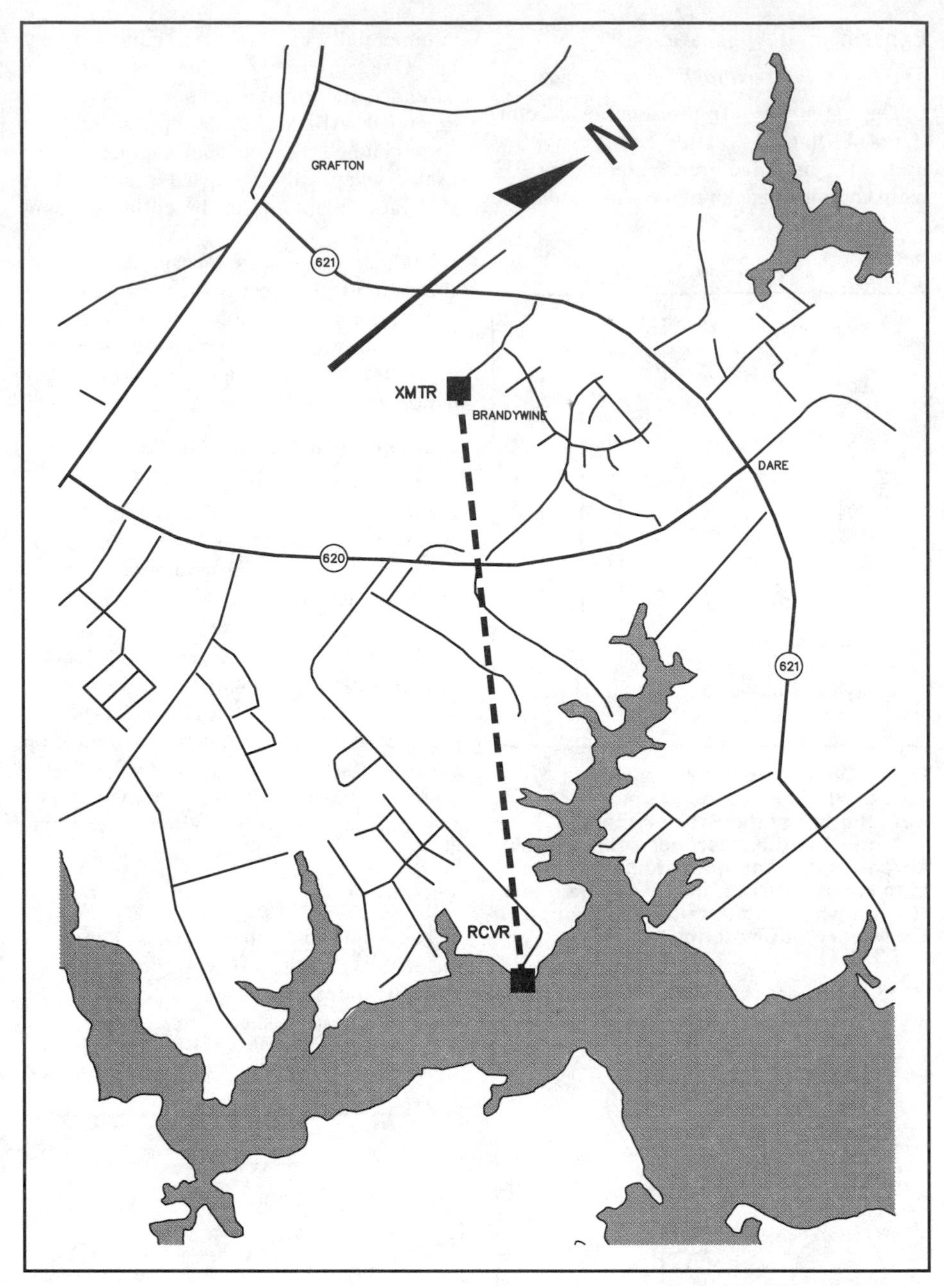

Fig 4—Test range in York County, Virginia; the distance from the transmit site to the receiver site is 1.6 mi.

Results and Discussion

The test results are divided into two basic groups. The first group consists of the short antennas, including the basic rubber duck (labeled #1) used as the baseline. The collapsed or duck versions of the telescoping antennas are labeled #2, 3, and 4. The second group consists of the long antennas, including the λ/4 whip, antenna #5, and the telescoping antennas, designated as #6, 7, and 8.

Short or "Duck" Antennas

Basic Influences: Fig 5 presents the behavior of the helical coil duck reference antenna used as the baseline for three test conditions with a tilt angle of 0° (antenna is vertical). The baseline antenna configuration data taken from different test runs illustrates clearly the gross asymmetry of the radiation pattern, showing a front-to-back ratio of more than 5 dB. Adding a stiff wire ground plane (4 wires of λ/4 spaced 90°) resulted in an almost symmetrical radiation pattern, simultaneous with a substantial decrease in forward signal level. With no ground plane, moving the operator a distance of about 5 λ laterally resulted in a near-circular radiation pattern of even greater strength than the best of the simulated hand-held configurations. This almost-circular pattern also indicates the high quality of the test setup and lends validity to the test results. The small remaining asymmetry is probably a result of the nonsymmetrical installation of the antenna on the HT itself, since it is mounted close to one side. Since this test was run with a short helical antenna, this relative level of 6 to 7 dB is probably the best attainable with this type of antenna under the most favorable antenna match conditions.

Some of the effects of tilt angle (set to 45°) are shown in **Fig 6** for the same three configurations. With the operator facing away from the receiving station (θ=180°), the signal strength is now almost equal to the case of the operator facing the receiving station, at an azimuth angle of 0°. The circu-

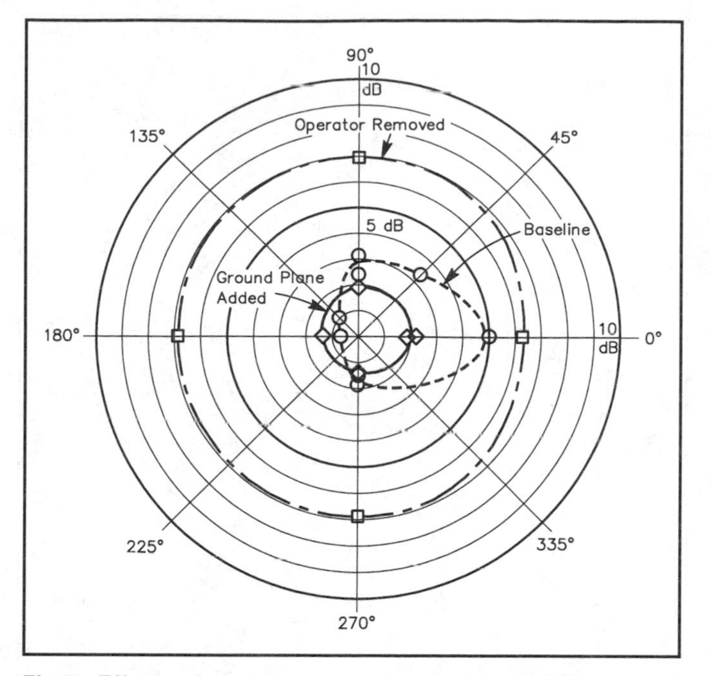

Fig 5—Effects of operator removal and ground plane addition on signal strength for antenna #1 (the baseline configuration), tilt = 0°.

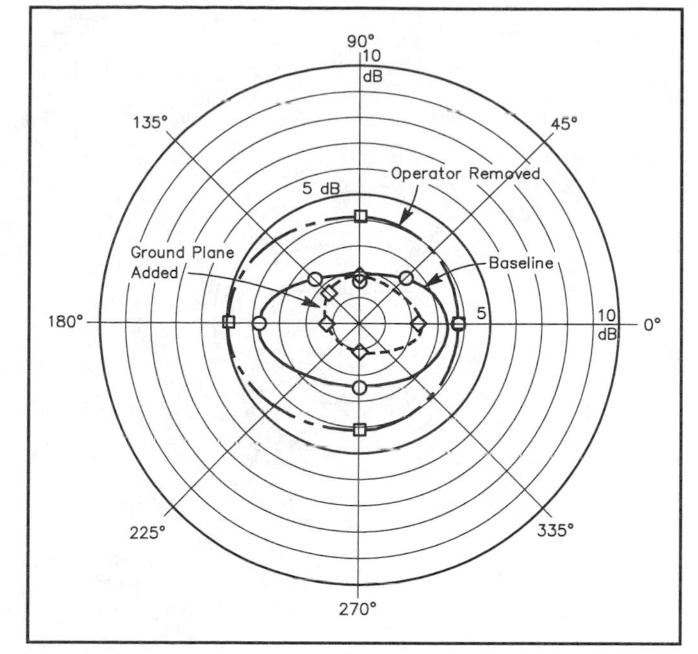

Fig 6—Effect of operator removal and ground plane addition on signal strength for antenna #1 with the tilt increased to 45°.

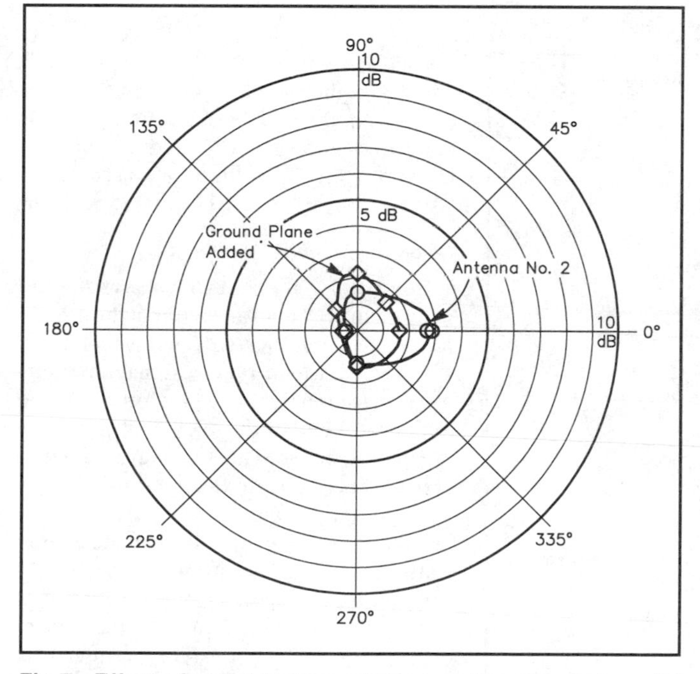

Fig 7—Effect of ground plane addition on the signal strength for antenna #2, tilt = 0°.

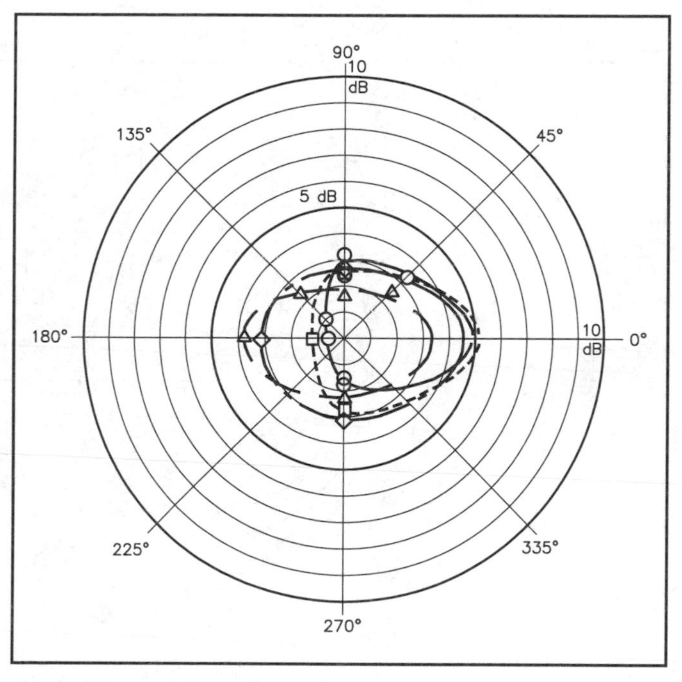

Fig 8—Effect of tilt angle on the signal strength for antenna #1, the baseline configuration.

lar pattern is distorted when the ground plane was added at this tilt angle.

When the operator was removed, and with no ground plane, a slightly distorted but still very strong signal was measured at all azimuth angles. This distortion was probably the result of the ground reflections coupled with the vertical plane lobe characteristics. These combined effects would be very difficult to express mathematically. Notice the large improvement in the signal strength, when tilted and facing away from the receiving station (see Fig 5).

Antenna Comparisons: **Fig 7** shows one of the telescoping antennas in the collapsed condition (so-called "duck" configuration) and with the ground plane added. Besides small level changes, there is a marked similarity of these normal HT configuration results compared with the helical-coil baseline antenna shown in Fig 5. Therefore, it may be inferred that this shape is characteristic of all the short antennas, whether a short helical coil, or some collapsed version of a long antenna. The level differences probably only reflect how well the antenna is matched when collapsed.

Effects of tilt angle are presented for the

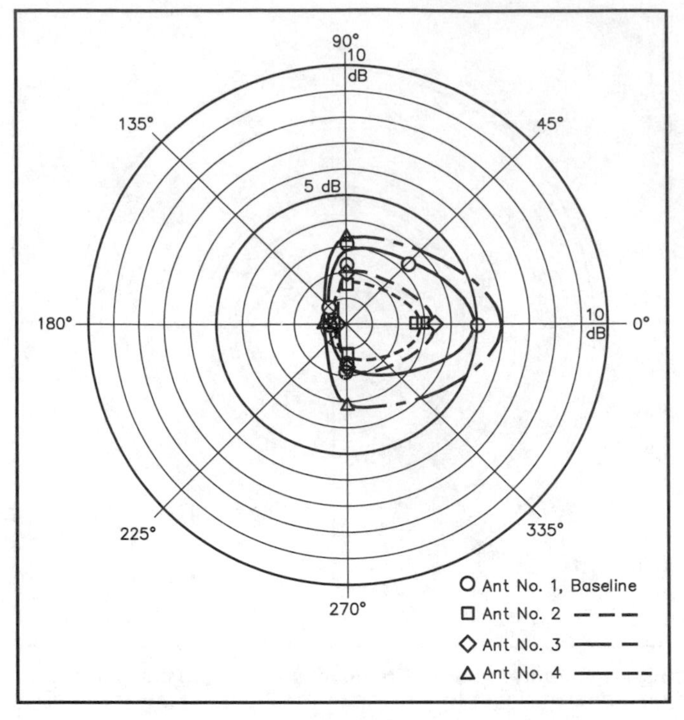

Fig 9—Comparison of the performance of four duck antennas, tilt = 0°.

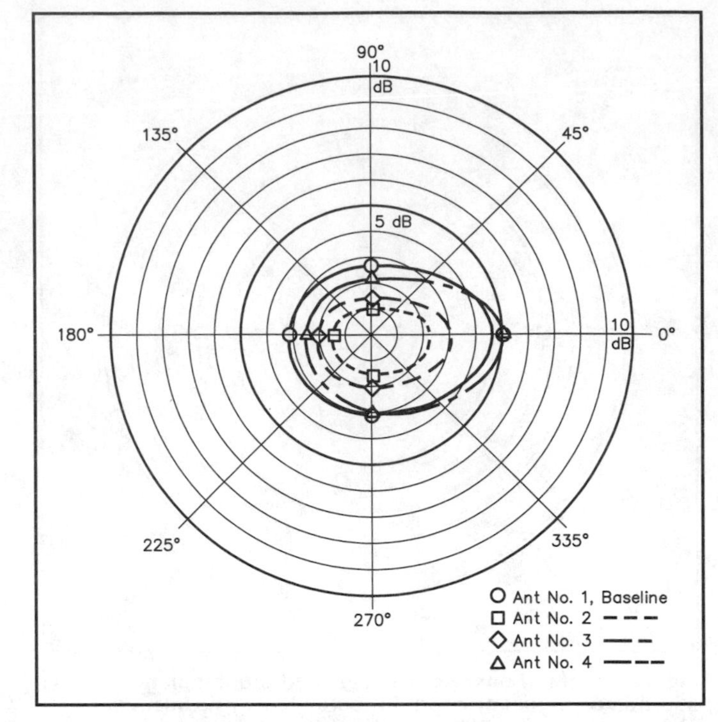

Fig 10—Comparison of the performance of four duck antennas, tilt = 30°.

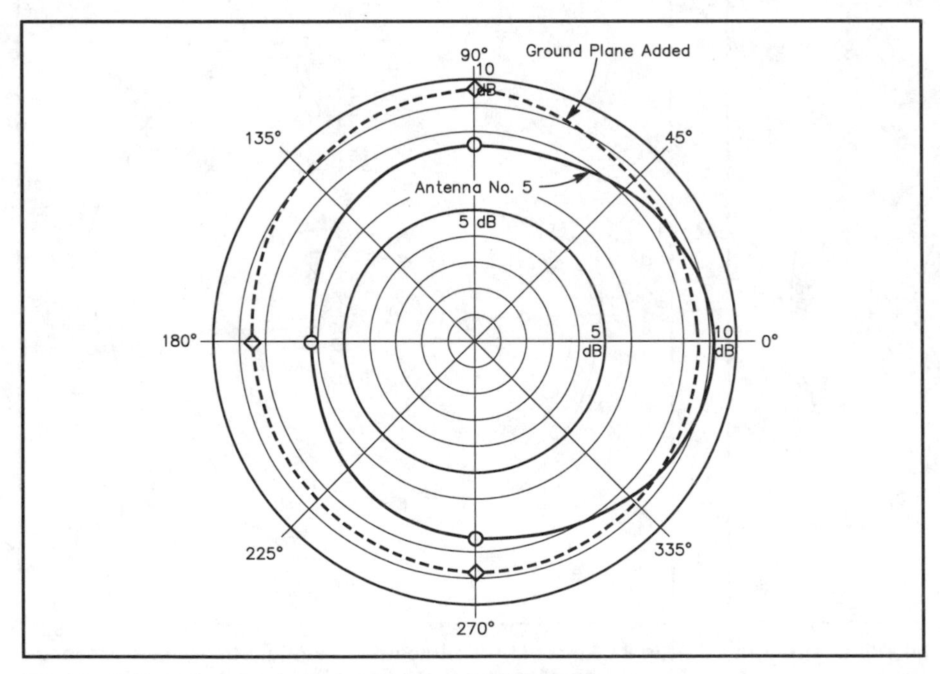

Fig 11—Effect of ground plane on signal strength with antenna #5, a λ/4 whip, tilt = 0°.

baseline duck, **Fig 8**, from the vertical (Ø=0°) to tilt angle of 45° (away from the operator). The progressive changes, especially at azimuth angle of 0° (facing the receiving station) and at 180° (facing away from the receiving station) are evident. The most rapid changes appear to occur between about 15° and 30° tilt angles.

The four "duck" configurations are compared in **Fig 9** at 0° tilt and in **Fig 10** at 30° tilt angle. The tilt angle was selected as being near the optimum, but on the high side in terms of radiation symmetry. The characteristics are all very similar except for the effect of mismatch. In all cases the forward radiation is several times greater than the rearward radiation when the antenna is vertical.

From these data it is also possible to show one design (antenna #4) has achieved a slightly higher signal level than the other designs (in the collapsed configuration) as compared with the baseline duck antenna. However, this slight difference has not been without other effects, to be pointed out later.

Long or Extended Antennas

Basic Influences: **Fig 11** shows the λ/4 whip on the HT with and without the ground plane made of four stiff wires λ/4 long, spaced 90°. Much higher signal strengths were measured for this λ/4 whip compared to those for any of the duck configurations previously shown. The signal levels are several decibels higher in all directions. For this antenna, only a moderate signal decrease results from facing away from the receiving station (θ=180°). Despite all that can be said concerning the antenna portion that does most of the radiating because of the current distribution, the upper part of the antenna is now well above the operator's head. The large, directional, operator-presence losses associated with short antennas shown earlier have now disappeared. The signal strengths measured with the λ/4 whip are in the same range as the values shown in Fig 5 for the baseline duck antenna when the operator was removed from the transmitter.

Antenna Comparisons: For antennas vertical (ϕ=0°), **Fig 12** presents a comparison of the performance of all of the long antennas referenced to the baseline duck. For this, and subsequent figures, the scale has been changed by a factor of two. If a wide paint-

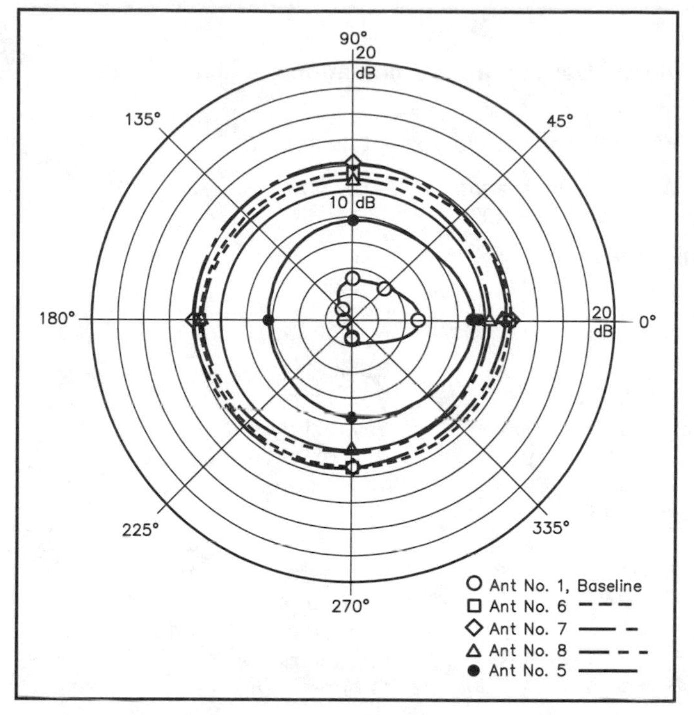

Fig 12—Comparison of the performance of the long antennas with the baseline duck and antenna #5, tilt = 0°.

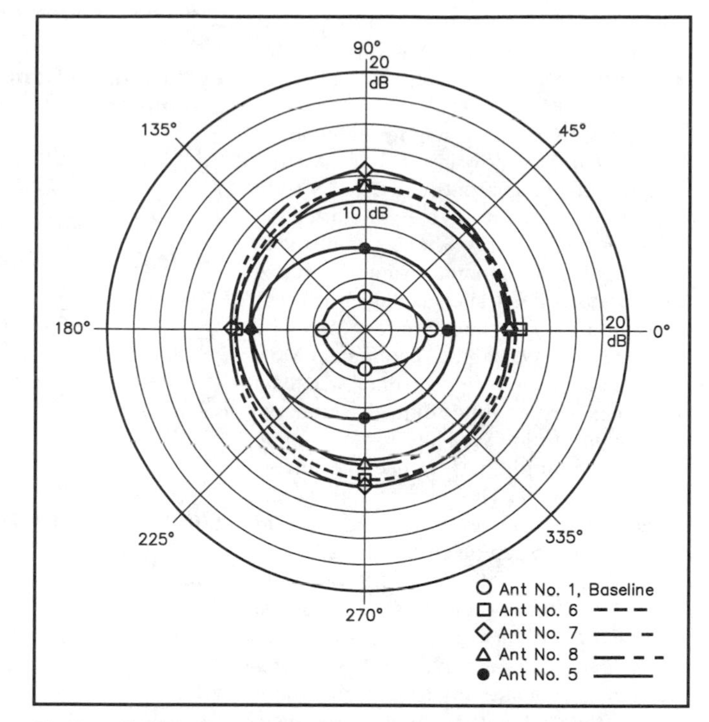

Fig 13—Comparison of the performance of the extended antennas with the baseline duck and antenna #5, tilt = 30°.

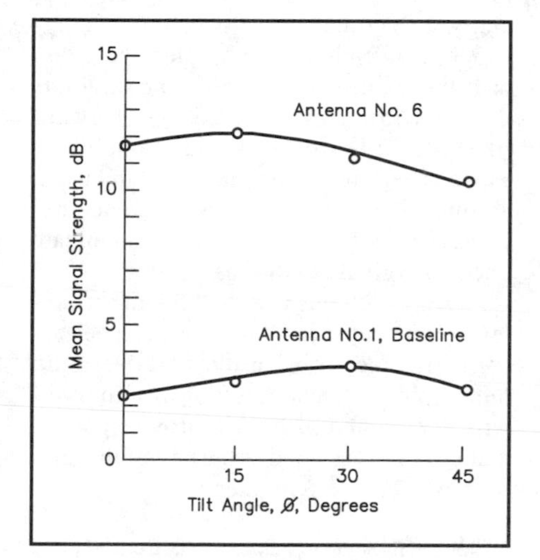

Fig 14—Variation of Mean Signal Strength, MSS, with tilt angle, antennas #1 and #6.

brush line, nearly circular in shape, were drawn, it would encompass the signal strength characteristics of all the extended versions of the telescoping antennas (whether $^1/_2\,\lambda$ or $^5/_8\,\lambda$). None of these extended antennas can be judged superior. As should be expected, they are all a little better than the $\lambda/4$ whip, antenna #5.

Fig 13 is a similar comparison of the long antennas with the baseline duck, but at a tilt angle of 30°. Once again, no one antenna is noticeably better.

Averaged Comparisons

Fig 14 is a comparison of the reference baseline duck configuration and a telescoping $\lambda/2$ antenna. The Mean Signal Strength, MSS, is plotted against tilt. MSS is an average of the measured signal strength values through the azimuth range of 0° to 360:

$$\text{MSS} = \int_0^1 (\text{dB})\ d(\theta / 360)$$

where dB is the measured signal strength at each azimuth angle.

The results show the MSS for a short duck-type antenna will average about 25 to 30% higher if the antenna is tilted to some angle probably, not less than 20° nor more than 35°. This suggests the natural attitude for all-around best communications will not be far different from the tilt angle at which many operators normally hold their HTs. A conscious effort at tilting has already been clearly shown (see Fig 8) to be advantageous if reception behind the operator is needed.

For long antennas, tilting a few degrees appears to offer only a minor benefit. More importantly, too much tilt in this situation is bound to hurt, especially if it is more than about 15°. This is not surprising, because the combination of vertical plane signal lobes and ground reflections almost always interfere.

Finally, **Table 1** provides a comparison of the incremental value of the MSS. The incremental MSS is simply the calculated MSS referenced to the baseline configuration. Since a duck-type antenna is usually the starting point for the user, the results are shown with the duck performance as the reference.

Using this method of comparison configurations 2 and 3 (collapsed telescoping antennas) appear about 1-dB poorer than the baseline (incremental MSS −1.2 and −0.9). These same antennas, fully extended as in configurations 6 and 7, show overall improvements averaging about 10 dB (11.8 − 1.8). This amounts to an incremental MSS of about 9 dB over the baseline—a substantial increase. However configuration 8 provided a gain of only 7.5 dB (10.8 − 3.3) in its fully extended condition. This was probably due to an attempt to obtain a design in the collapsed configuration (antenna #4) that would be better than the baseline duck. For configuration 8 the net incremental MSS is about 8 dB, or about 1 dB less than the other two telescoping antennas. Of course, the difference in the ability of the receiving station to hear is barely noticeable for gains of either 8 or 9 dB, so all of the three telescoping antennas must be considered to exhibit about the same overall performance.

Configurations 9 through 14 show some other effects. The presence of the human body, configuration 9, degrades the MSS by about 4 dB (2.7−6.7 = − 4.0). Tilting this arrangement 45°, configuration 10, resulted in severe ground reflection interference. Thus, the gain of removing the operator decreased from 4.0 to a mere 1.5 dB. The addition of ground planes to short antennas always degraded the MSS. For the helical coil duck, either vertical or tilted 45° (configurations 11 and 12), the incremental decrease in MSS was greater than 1 dB, or more than 30%.

For the collapsed telescoping antenna, its MSS was already down by 1.2 dB. Adding the ground plane reduced the MSS only slightly (0.1 dB), but severely distorted the radiation pattern. Only when a ground plane was added to an antenna extending above the operator's head, such as the λ/4 whip of configuration 14, could improvements be shown. For this case, the incremental MSS is about 6, and the received signal would be four times the magnitude of the baseline antenna, configuration 1. Thus, for this simple antenna, an output of 1 W becomes effectively about 4 W. This may well make the difference in whether or not satisfactory communication can be established.

All signal strength and MSS values in Table 1 and throughout this report are in relative dB and not with respect to any absolute scale. The authors, in hindsight, regret they did not attempt to tie the measured signal strength levels to an absolute scale, such as a reference λ/2 vertical dipole.

Table 1

Performance Comparisons Derived from Calculations of Mean Signal Strength (MSS)

All configurations are with antenna vertical except as noted.

Config Number	*Antenna Number*	*MSS dB*	*Incremental MSS, dB*	*Notations*
Basic Antenna Configurations				
1	1	2.7	+0.0	Baseline
2	2	1.5	−1.2	Duck
3	3	1.8	−0.9	Duck
4	4	3.3	+0.6	Duck
5	5	7.6	+4.9	1/4-λ whip
6	6	11.6	+8.9	1/2-λ extended
7	7	11.8	+9.1	5/8-λ extended
8	8	10.8	+8.1	5/8-λ extended
Other Antenna Configurations				
9	1	6.7	+4.0	Operator removed
10	1	4.2	+1.5	No operator, $\phi = 45°$
11	1	1.7	−1.0	Ground plane
12	1	1.5	−1.2	Ground plane, $\phi = 45°$
13	2	1.4	−1.3	Ground plane
14	5	8.9	+6.2	Ground plane

Concluding Remarks

A carefully controlled experimental investigation has been performed to measure the performance of several HT antenna configurations, including three telescoping antennas. Test parameters included azimuth angle and tilt angle. The following are the principal results:

1. Test range and test mount considerations are important in order to obtain high quality data.

2. Azimuth angle is especially important when using short antennas such as ducks, or collapsed telescoping antennas.

3. Tilt angle produces large, often favorable, signal changes for all short antennas.

4. None of the telescoping antennas tested proved markedly superior in their collapsed configurations.

5. Addition of a 4-element stiff wire ground plane to a short, or duck antenna produced generally degraded signal strength patterns.

6. Operator body influences are very large, signal-degrading, and position-sensitive, particularly for short antennas.

7. Azimuth angle is relatively unimportant for long antennas extending above the operator's head, such as λ/4 or longer antennas.

8. Tilt angle is relatively unimportant for long antennas extending above the operator's head, such as λ/4 or longer.

9. None of the telescoping antennas proved markedly superior in their extended configurations; in fact, all gave similar results.

In addition, these results indicate the following recommendations for the user:

1. For station-to-station communication, always face the other station, if its location is known.

2. While operating simplex in the field with a rubber duck antenna, if one needs to blank a particular station, turn 180° away from that station. The human body will block most of the interfering signal, especially if the rubber duck is held vertical.

3. For multiple station simplex operation, or if the location of the receiving station is unknown to the operator, moderate forward tilt angles of 15° to 30° should be employed, especially if duck antennas are being used. Tilting may well make the difference between success and failure to communicate under marginal conditions;

4. An excellent, low-cost, compact, antenna is a simple λ/4 whip. One can be easily fabricated from an old portable radio telescoping antenna. Addition of a simple 4-wire λ/4 ground plane will provide performance almost as good as the best of the telescoping λ/2 or 5/8 λ antennas.

[1] *The ARRL Antenna Book*, 17th Edition (The American Radio Relay League, 1994) pp 27-43 to 27-48.

from *The ARRL Antenna Compendium, Vol 2*

The Half-Wave Handie Antenna

By Ken L. Stuart, W3VVN
1235 Hill Creek Rd
Pasadena, MD 21122

The garden variety 5/8-λ whip is not the ultimate choice for use on a hand-held transceiver. For effective operation, a 5/8-λ antenna must work against a good ground plane, which a hand-held radio certainly is not. Also, the additional 1/8 λ over a half-wave actually produces an out-of-phase component which creates additional lobes and gives a vertical angle to the radiation pattern. (See the treatise on 5/8-λ antennas in *The ARRL Antenna Compendium Volume 1*.[1]) The main advantage of the half-wave is the gain over a 1/4-λ whip, and it is easy to load.

The Half-Wave Handie Antenna is an improvement over the 5/8-λ whip. Actually, the half-wave is the perfect antenna for producing the desired doughnut-shaped radiation pattern. And by end feeding it, the need for a ground system is essentially eliminated. It is also slightly shorter in length than a 5/8-λ, which reduces strain on the hand-held transceiver coax connector. A commercially manufactured version of this antenna has been available for several years and has become very popular (the AEA Hot Rod).

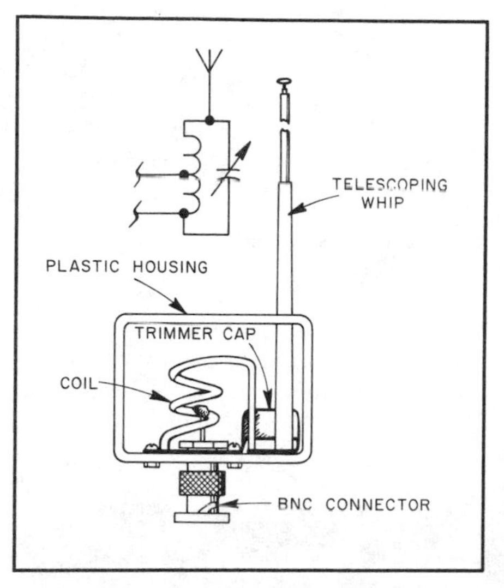

Fig 1—Construction details of the Half-Wave Handie Antenna. The enclosure is formed from a strip of thick plastic.

Whip—Radio Shack 15-232 (page 101 in the 1988 catalog).
Trimmer capacitor—6 to 50 pF, Radio Shack 272-1340 or Arco 404.
BNC connector—UG-260 or UG-88.
Coil—3 turns no. 18 bus wire, ½-inch diameter, ½-inch length. Tap approximately one turn from ground. (Adjust for minimum SWR.)

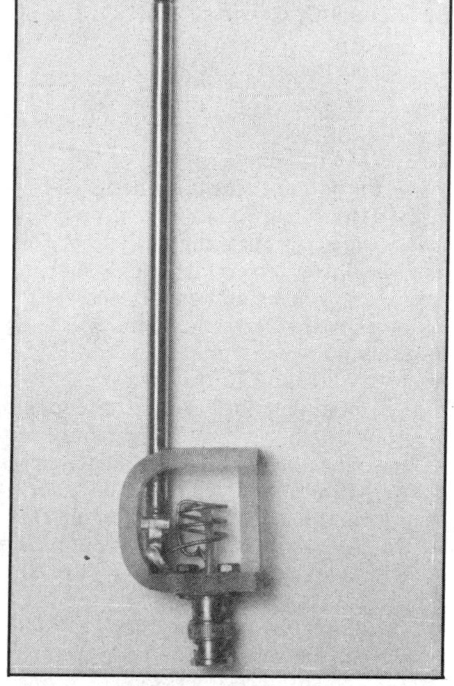

Fig 2—The trimmer capacitor is on the left side of the coil and directly in front of the antenna. (*Photo courtesy of Dean Alley, NS3V*)

Construction

Being of Scottish descent, I decided that I could build my own version of one of these antennas for less money, so I took several pieces of plastic and wire, a BNC connector, a trimmer capacitor and a Radio Shack replacement whip antenna element and fashioned my own junk-box special. It has operated on my hand-held transceiver for several years, and produces excellent results. In fact, in tests of this antenna, a 1/4-λ whip and a 5/8-λ whip, this antenna "whipped" them all.

This antenna works on the principle that a ½-λ dipole is an excellent radiating element, and by feeding the element at one end where the impedance is highest, we can eliminate the need for a ground plane. One common utilization of this principle is the popular "J-pole" antenna, where the radiating element is actually a ½-λ element fed at the end by a 1/4-λ matching stub. (The J could be used on a hand-held transceiver, and would work very well, but an antenna length of almost 5 feet atop a hand-held unit would be impractical.)

Electrically, the ½-λ antenna consists of a 39-inch collapsible whip with a parallel-tuned matching network connected between it and the hand-held transceiver ground. RF from the radio is applied to a tap on the coil of the matching network. Tuning of the antenna is accomplished by moving the tap feed point and adjusting the trimmer capacitor to achieve a minimum SWR.

Figs 1 and 2 will aid in understanding the following construction steps. Mechanically, I formed a strip of thick plastic into a four-sided box shape with a pair of opposite sides open. (See August 1988 *QST* for detailed information on building enclosures.[2]) A piece of copper was drilled to accept the BNC-connector cable-clamp nut, which was dropped into the hole and its flange soldered to the copper. Four holes were drilled in the copper and the plastic housing for mounting the BNC. Mounting of the whip was done by drilling a snug clearance hole in the top of the housing and a small hole in the bottom to accept the whip mounting screw. The coil and trimmer capacitor were mounted simply by soldering them to the new BNC connector copper flange and a solder lug at the bottom of the whip. A piece of bus wire connects the center pin of the BNC to the tap point on the coil. It is soldered to the center pin of the BNC, the pin is inserted in the connector, and the pin and wire potted in place with epoxy cement or hot glue.

Operation

Although the antenna is properly matched only when the whip is extended to full length, I have used it on low power with the whip collapsed with no ill effects to the radio. I don't use high power with the whip collapsed, however. Hand-held transceivers as a breed are pretty bullet-proof, but I prefer not to abuse mine by trying to work into a mismatch at a full power setting.

This antenna has proved itself time and again. It has seen use in numerous RACES and ARES drills and emergencies, and in public service events. It lives on my radio at all times. It is replaced by a duck only when communications will be needed over a very limited geographic area, when it will be more convenient to wear the hand-held transceiver on my belt and use a speaker mike.

[1]D. K. Reynolds, "The 5/8-Wavelength Antenna Mystique," *The ARRL Antenna Compendium, Volume 1,* pp 101-106.

[2]D. Kennedy, "Build It Yourself—With Plastic," *QST*, August 1988, pp 30-34.

from *The ARRL Antenna Compendium, Vol 2*

Portable 2-Meter Antenna

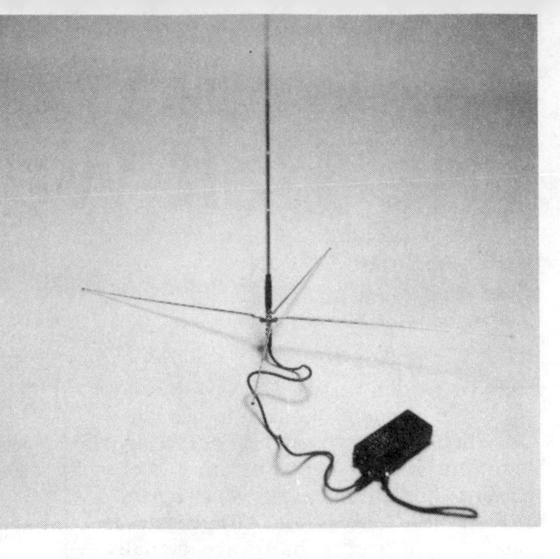

By Michael C. Crowe, VE7MCC
Box 316
Madeira Park, BC V0N 2H0

This portable antenna is designed for long-range communications on 2 meters while using small mobile or hand-held transceivers. It is light, simple, rugged and convenient to carry and erect. Although based on common principles and breaking no new technical ground, it has been very effective in the field.

It is best described as a telescoping 5/8-wavelength vertical whip above a ground plane of four folding telescoping quarter-wave radial whips. The theory came from material in older editions of *The ARRL Handbook*; my only contribution is in the simple folding construction to make it portable.

The ubiquitous "rubber duck" is convenient but has severe shortcomings when terrain and distance intervene. In trying different antenna configurations on my hand-held transceiver while searching for a VHF antenna for use in hiking and camping and also for emergency operations, several things became apparent. I noticed that, especially in transmitting, a ground plane gave considerable assistance to most omnidirectional vertical antennas and seemed also to improve the SWR slightly. The height of at least a quarter wave, or even better, half or 5/8 wave, helped transmissions when compared to a compact helical or "rubber duck" antenna. It came down to a decision between loaded 5/8- or 1/2-wave verticals. The 5/8 tested out better, and my efforts then concentrated on building such a ground-plane antenna in a form practical for field work.

After trying various configurations, I came upon the idea to incorporate a ground plane of telescoping whips structurally into the antenna base by affixing them to a sturdy washer mounted there. After that, the actual prototype construction took only an hour, utilizing a second-hand telescoping 5/8-wave whip. The SWR across the 2-meter band is less than 1.5 to 1 (it can be adjusted to no more than 2.0 to 1 on the marine band, so is useful on boats, also). In use, the SWR varies with antenna tilt (see later text), closeness to vegetation, dampness, etc, at any one frequency. So if you are not satisfied with the SWR, try different conditions.

Construction

As Figs 1 and 2 show, construction centers on an ordinary steel washer, about 2½ mm thick and 4 cm in diameter. Four equally spaced holes (sufficient to take a 3-mm bolt) are drilled around its circumference. The washer is then mounted on a female-to-female BNC connector (mine is an Amphenol part) designed for panel mounting so that its body is threaded. Tighten the washer around the body of the connector using a large lock washer and the two nuts supplied (if none are, scrounge a couple). It is also possible to find BNC connectors with pre-drilled flanges on their bodies for chassis mounting, but these appear too delicate for portable use.

Fig 1—The ground-plane portion of the portable 2-meter antenna. Construction centers on an ordinary steel washer, about 2½ mm thick and 4 cm in diameter.

Several suppliers sell replacement telescoping antennas for portable AM radios which have a flattened mounting portion at one end. Most already have a small hole drilled in the flat portion and an elbow where the whip itself is screwed. Find four of these (they are often sold by Radio Shack, among others, in packages of four for about $10; I wonder who otherwise buys packages of them...). One is mounted on each hole in the large washer, with

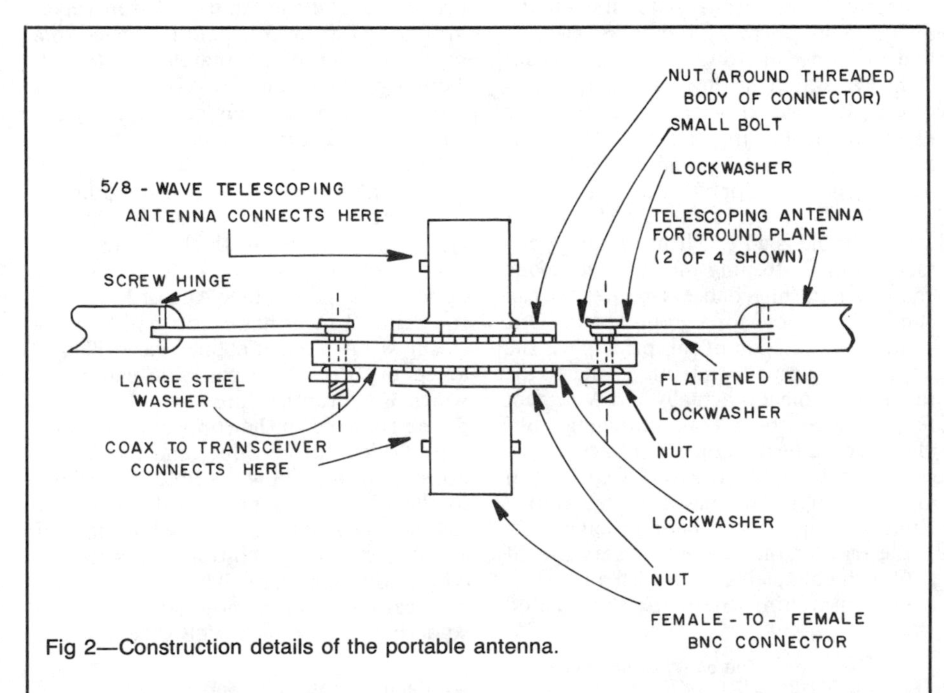

Fig 2—Construction details of the portable antenna.

a lock washer on each side and other small washers as needed, to be tightly mounted by the 3-mm bolts and nuts (or such other size as is convenient).

Attach a 5/8-wave base-loaded telescoping whip to the upper BNC connector, and a length of coax cable (mine is 1 meter; somewhat longer would be better) with male BNC connectors at each end to the lower connector. The other coax end then connects to the BNC connector on your hand-held or mobile 2-meter transceiver. That is all there is to it.

Operation

The antenna is basically omnidirectional, and the best results (please do not ask why) are achieved by fully extending all whips and tilting the vertical whip slightly *away* from the desired direction of transmission. Having the horizontal whips bunched on one side further favors transmitting in that direction. For reception, the ground-plane whips can be left telescoped without significant sensitivity loss, but for maximum transmitting efficiency, their extension is essential.

This forms an antenna 130 cm high and 95 cm wide when fully extended, but folds and telescopes to a package only 5 cm in diameter by 24 cm long. It weighs less than half a kilogram, and easily packs in its folded state by wrapping an elastic band around it. Over a rubber duck, the antenna has a theoretical advantage of about 8 dB (I do not have the equipment to measure this, but my results support it). The addition of extra horizontal whips in the ground-plane array seems to have no discernible effect, nor does the use of longer horizontal whips. Do not worry about droop; even if bent at 45 degrees, they are still effective.

Its first use was in the mountains near Vancouver in the fall of 1987. In a clearing (definitely avoid overhead vegetation) in a valley behind a 2000-meter mountain ridge, I was able to make contacts through two repeaters on apartment buildings in Vancouver, about 140 straight-line kilometers away. My signal was adequate to make telephone calls on the autopatch on one of them. Contacts indicated my signal strength to be sufficient for only partial quieting but fully intelligible.

More recently, during an emergency-preparedness exercise, I was able to access repeaters that more powerful mobile transceivers in vehicles with mag-mounted whips could not. Several more of these antennas have been built by local hams in the Vancouver area.

from April 1982 *QST* (Technical Correspondence)

DECOUPLE VHF VERTICALS

☐ Improper decoupling of transmission lines from vertical antennas mounted above ground can destroy the predicted radiation patterns. I built W1FB's 5/8-wave antenna (June 1979 *QST*) for 2-meter fm service and erected it a few feet above my roof. The performance did not meet my expectations — it was scarcely better than a quarter-wave ground-plane.[1]

After researching the subject, I concluded that current was flowing on the outer conductor of the coaxial feed line. This current produced fields that interacted with the antenna fields, resulting in an upward deflection of the radiation lobe, which meant degraded performance.

I built the decoupling network shown in Fig. 1 and added it to the original antenna. Now the feed line goes through the TV mast and attaches to the bottom of the antenna. It is essential that a good mechanical and electrical connection be made between the braid (antenna ground) and mast top. Instead of attaching the radials to the loading coil base, I mounted them on the mast with a hose clamp at a point 5/8-wavelength below the loading coil. In a similar fashion, I affixed a second set 1/4-wavelength below the first.

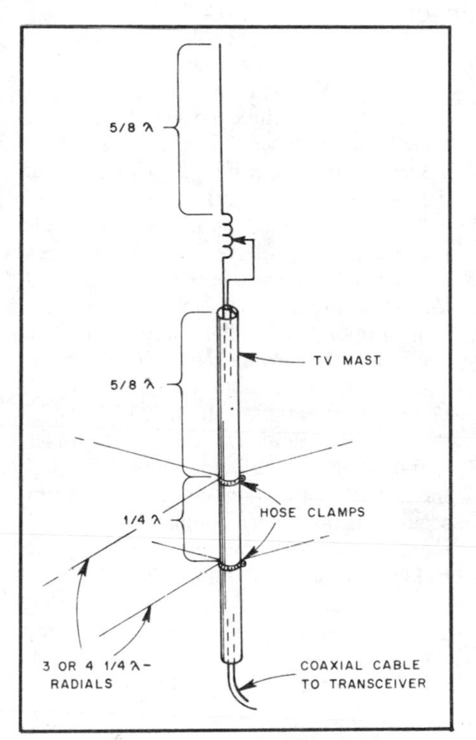

Fig. 1 — The supporting mast becomes part of the decoupling network.

Although I expected some improvement in performance, I was pleasantly surprised at the difference in received signal strength. A station located approximately 10 miles (16 km) away acts as a control for a MARS net on 143.99 MHz (simplex). Before adding the decoupling circuit, I could barely copy him. After the addition, his signal consistently fully quiets the receiver and pins the S meter on my Heath VF-7401. He reports a similar improvement in my transmitted signal strength.

In the present configuration, the antenna would be described as an extended double Zepp. The name is not important, but improved performance is. Because the fields produced by feed-line radiation vary from one installation to another, the amount of improvement after decoupling will also vary. If you are not getting the performance from your vhf vertical that you expect, you may want to add a decoupling system. The improvement may surprise you. — *Peter O'Dell, KB1N*

[1]The W1FB design was developed for mobile use, which provided a ground plane by virtue of the car body. The feed line was therefore contained within the ground system.

from February 1980 *QST*

A VHF-UHF 3-Band Mobile Antenna

Three bands — 144, 220 and 440 — on one stick sound interesting? This antenna might allow you to condense that stainless-steel and plastic jungle atop your auto onto a single pole.

By J. L. Harris,* WD4KGD

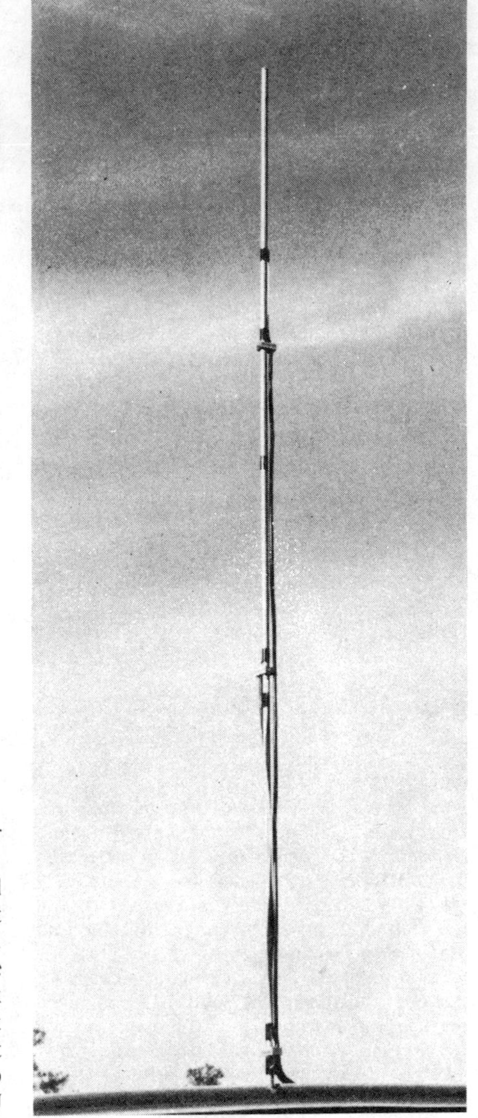

The three-band antenna system mounted atop a pickup truck. *(photo by WD4FNS)*

In looking for a mobile antenna system for my Drake UV-3, I rejected the notion of one broadband antenna such as the discone because of band-switching problems not to mention its somewhat busy appearance. I also rejected the idea of three separate whips which I felt would give the relatively small roof area of my pickup truck a cluttered look. Three separate antennas confined to so small a space would also cast "shadows" on the vertical patterns of one another. In order to take full advantage of the three antenna terminals on the UV-3, I needed three separate antennas, but I wanted an omnidirectional pattern with no "holes."

The solution I chose was to use three stub-fed verticals on one whip. The stub-fed vertical, or J antenna, consists of a basic half-wave radiator end fed through a quarter-wave stub. This stub serves as an impedance transformer. It transforms the high impedance of the half-wave radiator to that of the low-impedance coaxial line. Few antennas lend themselves to omnidirectional patterns and ease of matching to coaxial line as well as the stub-fed vertical.

Construction

My approach is cheap, novel and effective and uses only four basic parts except for the coaxial lines: the whip and three easily fabricated blocks. These materials are available at most hardware or hobby stores. The whip is one piece of 3/8-inch (9.5-mm) aluminum tubing 60 inches (152 mm) in length. Be sure that the piece you select is straight and free of nicks or dents.

Overall construction is shown in Fig. 1. The three stub blocks are made from 3/8-inch (9.5-mm) aluminum stock. Refer to Fig. 2 and saw three blocks 3/8 × 5/8 × 1-1/8 inches (9.5 × 15.9 × 28.6 mm). Drill a 3/8-inch (9.5-mm) hole as shown so that the piece will slip over the mast. Tap a no. 6-32 hole into the 3/8-inch (9.5-mm) hole just drilled for a setscrew to hold the block in place. The third hole is used to connect the braid of the coaxial cable to the mast. It is at this point where the quarter-wave stub begins and the feed line ends. For RG-58/U and similar size cable use a 13/64-inch (5.2-mm) drill and tap the hole with 1/4-20 thread. For RG-8/U, use a 25/64-inch (9.9-mm) or "X" drill and tap with 7/16-20 thread. Prepare the coaxial cables by separating the center conductors from the remainder of the cable to the lengths given in Fig. 1. Cut off all but 3/8 inch (9.5 mm) of the braid and fold this back over the jacket. These sections can be threaded into the tapped holes. The blocks can then be mounted to the whip as in Fig. 1.

Matching

As mentioned earlier, the quarter-wave stub is an impedance transformer. The spacing between the coaxial cable center conductor and the whip (dimension "A" in Fig. 1) determines the impedance of this section and consequently the match to 50-ohm line. Using an SWR indicator, determine the optimum spacing "A." This dimension can vary greatly depending on the size of the cable and its dielectric material. Once I determined the correct spacing, I stood off the center conductor from the main support with small styrofoam blocks. Electrical tape was used to hold the quarter-wave section and styrofoam block to the main support.

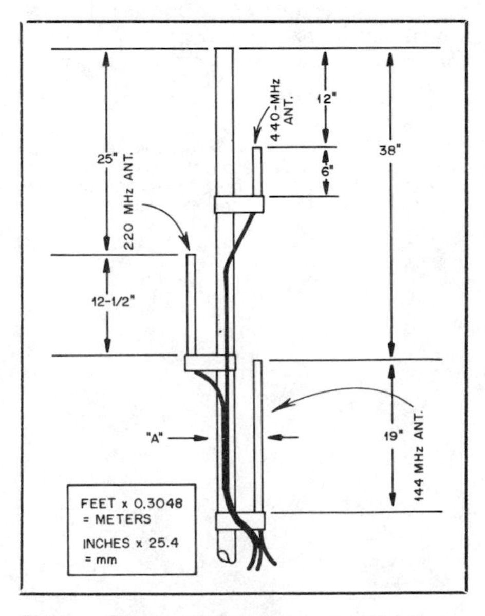

Fig. 1 — Construction dimensions of the three-band antenna. Cables should be routed and taped as shown.

*Rte. 5, Box 39, Henderson, NC 27536

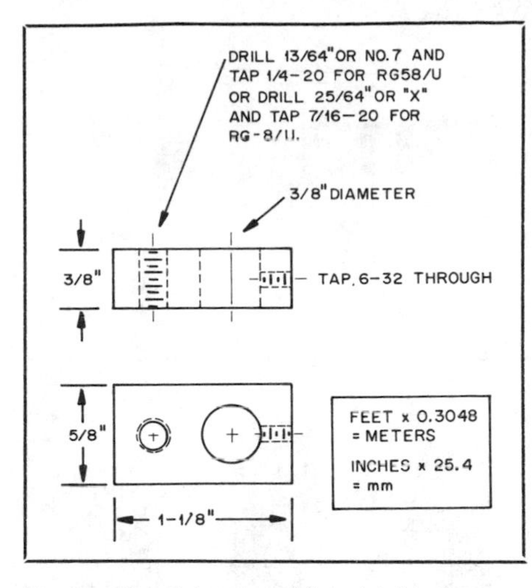

Fig. 2 — Detail drawing of the stub blocks used to connect and support and the quarter-wave sections.

The cables from the 440- and 220-MHz antennas should be routed as shown in Fig. 1 on opposite sides of the main support and away from other stubs.

The assembly is finished by taping all cables in place and coating the stub blocks with clear acrylic spray to prevent moisture from entering the cables. Although this antenna system is intended for mobile use and is constructed for this purpose, it should not be overlooked as a base station system. Just add 6 meters and you've got a 4-band array! E-plane patterns for the three bands are shown in Fig. 3.

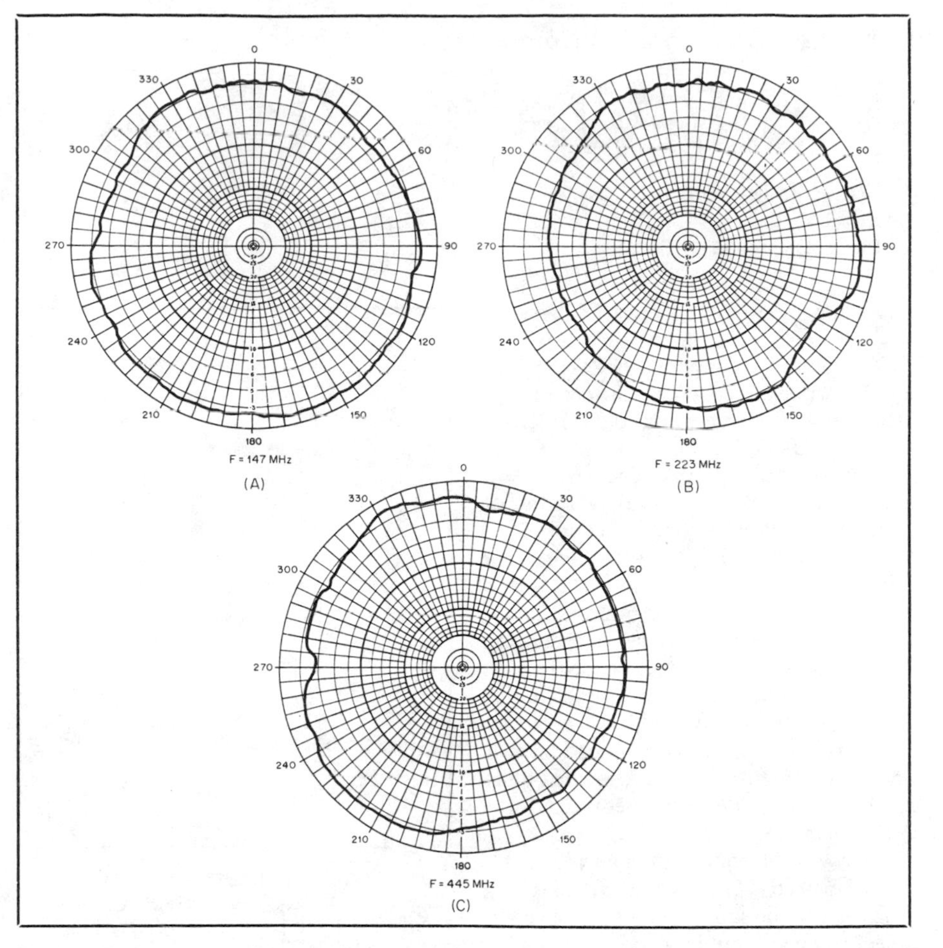

Fig. 3 — E-plane patterns for the three-band antenna. The patterns at A, B and C, respectively, are measured responses for 147, 223 and 445 MHz.

A 5/8-Wave VHF Antenna

By Don Norman, AF8B
41991 Emerson Court
Elyria, OH 44035

This antenna grew out of a series of experiments with feed-line decoupling. Ralph Turner, W8HXC, and I began exploring feed-line decoupling after Ralph discovered considerable RF on the feed line of a popular commercial 2-meter vertical antenna. The writer built a 5/8-λ antenna and began a series of experiments with feed-line decoupling.

Since this was a homemade antenna, the radials were attached to the mast with a homemade ring clamp and could be repositioned very easily. A series of tests and measurements proved to my satisfaction that 1/4-λ radials do not belong near the matching network on a 5/8-λ antenna. Quarter-wave radials positioned 3/8 λ below the matching network worked quite nicely and yielded excellent feed-line decoupling.

I believe that a 5/8-λ antenna with 1/4-λ radials placed 3/8 λ below the matching point acts in the same manner as the venerable extended double Zepp antenna. In fact, the antenna works well when the matching network is removed and it is center fed with 300-Ω balanced line. The balanced line must be led away from the antenna at right angles for more than 1 λ.

When the decoupling experiments were finished, there were a large number of odds and ends on hand and a search was begun for a design of an antenna that the average amateur could build with ordinary hand tools and hardware store and Radio Shack items. A sketch of the antenna is presented in Fig 1.

The radiator is cut from ¾-inch OD aluminum tubing and the supporting mast is a 1¼-inch OD television antenna mast. The matching rod and radials are made from hard-drawn aluminum clothesline wire. The center insulator is made from a pipe coupling for the sort of semiflexible plastic water pipe that is joined with molded plastic fittings and stainless steel hose clamps. Fig 2 shows three views of the plastic pipe fitting.

Fig 2A is a sketch of the coupling before anything is done, 2B is a cutaway of the coupling inside the mast, and 2C is the radiator inside the insulator. The ¾-inch aluminum radiator is a loose fit inside a 1-inch pipe coupling, and the 1-inch pipe coupling is a loose fit inside the TV antenna mast. These loose fits are tightened with shims cut from aluminum beverage cans. The radiator is installed in the insulator by inserting the tubing halfway through the coupling and drilling a hole and installing a self-tapping sheet-metal screw. The radials are attached to the mast with self-tapping sheet-metal screws.

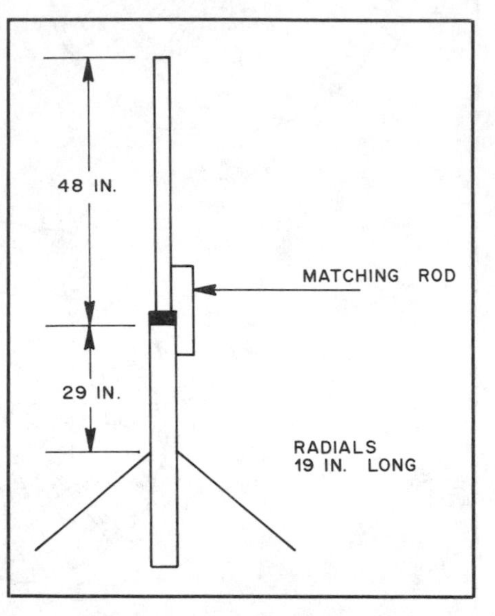

Fig 1—Diagram of the 147-MHz 5/8-λ antenna. It is designed to be built with readily available parts and ordinary hand tools.

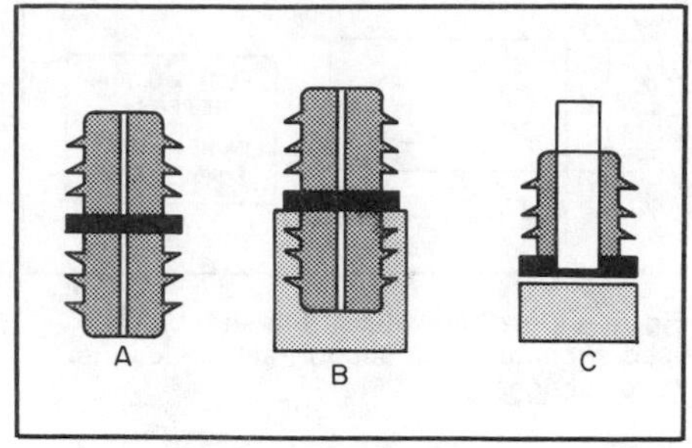

Fig 2—Three views of the plastic pipe fitting inside the center insulator.

Fig 3 is a sketch of the matching network of the 5/8-λ antenna. The matching rod is bent up from a 19-inch length of hard-drawn aluminum wire. Measure 2 inches from one end and make a right-angle bend. Make another right-angle bend 1 inch from that one. Measure 2 inches from the other end and make a right-angle bend. Make another right-angle bend 1¼ inches from the end. You should now have a U-shaped piece of wire with the U 1 inch deep at one end and ¾ inch deep at the other.

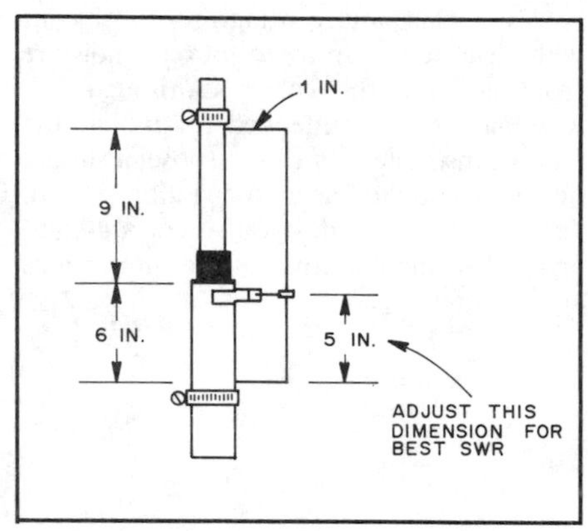

Fig 3—Detailed view of the matching network.

Drill a 3/8-inch hole through the mast and insulator between two of the mast attachment screws. Fish the coax through this hole. Attach the matching rod to the radiator and mast with stainless steel hose clamps according to the dimensions in Fig 3. (Be sure to file any paint or anodizing off the mast and radiator.) Ground the coax shield to the mast under one of the mast attachment screws. Attach the coax center conductor to the matching rod with a homemade clamp. Adjust the coax tap position on the matching rod for best SWR.

The same design works well at 220 MHz. A 220 antenna was designed, constructed

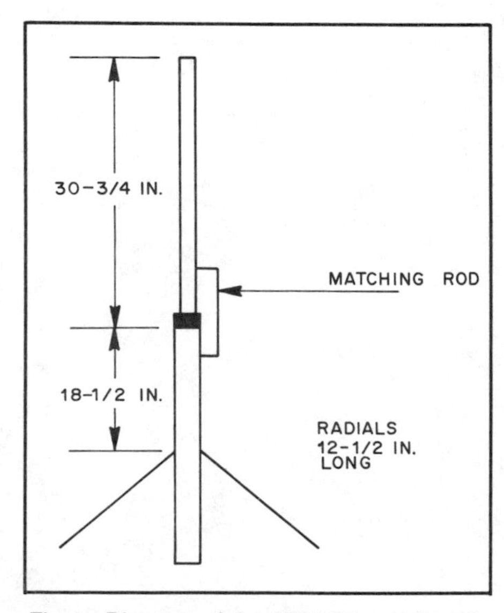

Fig 4—Diagram of the 220-MHz version of the antenna.

and tested in 1982, but only recently became popular locally. Fig 4 gives overall dimensions for the 220 antenna. Materials and construction of the 220-MHz version are the same as for the 147-MHz version; only the lengths are different.

Fig 5 is the matching detail for the 220-MHz antenna. The antenna may be built for center frequencies other than 147 and 220 MHz. Formulas are (all dimensions in inches, f = center frequency)

radiator—7056/f
radials—2793/f
radial attachment point below top of mast—4263/f
matching rod—2205/f
matching rod attachment point above top of mast—1323/f

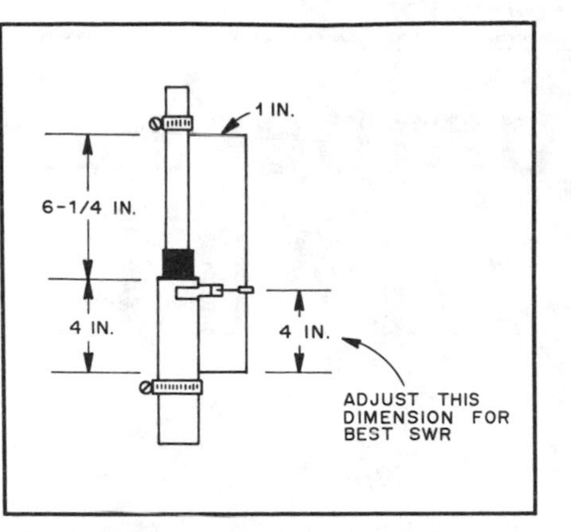

Fig 5—The matching network for the 220-MHz version.

The matching rod is spaced 1 inch from the radiator for both the 2-meter and 1¼-meter bands.

This antenna design works well, whichever frequency it is constructed for. We have found that matching is easier if the feed line is cut in multiples of a half wavelength at the most common operating frequency. A half wavelength in inches is determined by the equation

$$L \text{ (inches)} = 5904/f \times VF$$

where

f = frequency in MHz
VF = velocity factor of the particular cable

from January 1980 *QST* (Hints and Kinks)

A HARDLINE COAXIAL ANTENNA FOR 2 METERS

I ran across Phil Rand's antenna article in November 1951 *QST* while searching for data on coaxial antennas that could be made from obtainable materials and fabricated without the benefit of a machine shop. A similar antenna could be made from CATV Hardline cable. Often, odd length pieces of such cable are available from CATV companies.

I have corresponded with Phil about my idea (as shown in the accompanying diagram). He advises that there is no need to consider the velocity factor, inasmuch as the coaxial cable feeds only the top element. He also suggests that I keep the same ratio of mast-to-skirt diameters, stating that the skirt diameter should be 1-1/4 inches. Additionally he advises that the 19-inch dimensions are for the 144-MHz end of the 2-meter band. These measurements should be changed to 18 inches for the fm segment. — *Harry H. Heinrich, W9KPG, Green Bay, WI*

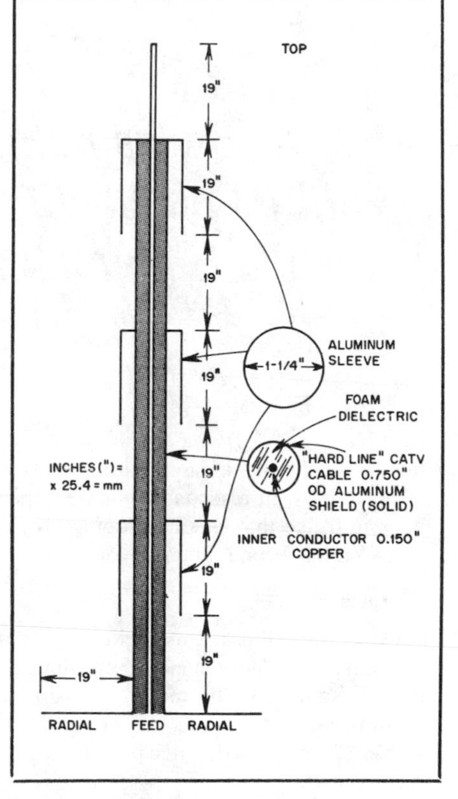

W9KPG suggests this antenna design for 2-meter operation. It is constructed from Hardline CATV coaxial cable. Measurements shown are for the 144-MHz end of the band. The measurements should be shortened to 18 inches for the fm segment.

A True Plumber's Delight for 2 Meters—An All-Copper J-Pole

By Michael P. Hood, KD8JB
2323 Jefferson Drive SE
Grand Rapids, MI 49507-3148

KD8JB was not happy with how his old J-Pole held up in the weather, so he made a much more rugged one.

A number of years ago, I built a J-Pole antenna from some scrap aluminum tubing and a 5-foot stick of TV mast.[1] The antenna tuned easily, and operated well for a few years until nature did it in. After taking it down, I found a number of problems not contemplated in the original construction, such as dissimilar metals and other types of corrosion, and deterioration of parts of the coaxial feed line. These problems are identified and explained in the accompanying sidebar.

With the shortcomings of the original J-Pole in mind, I set about planning for another, similar antenna, using parts commonly available from a single source. If you spend much time in hardware stores and home improvement places, you get a feel for what's available at a moment's notice. In my case, rigid copper tubing, fittings, and assorted hardware, came to my attention as the material of choice. See **Fig 1**. The entire assembly can be soldered together when copper is used, thus ensuring electrical integrity, and making the whole antenna weatherproof in the bargain.

I'll bet you could solder one of these together faster than using the nuts and bolts I did in my previous design. This antenna can be easily used for ARES/RACES groups that spot antennas around for emergency use, since it requires little, if any, maintenance during its lifetime.

The J-Pole will take about an hour or so out of your day to build and tune, making a great antenna for a VHF base station.

No special hardware or machined parts are used in this antenna, nor are insulating materials needed, since the antenna is always at dc ground. Best of all, even if the parts aren't on sale, the antenna can be built for less than $15. If you only build one antenna, you'll have enough tubing left over to make most of a second antenna, or perhaps to finish that small plumbing project the XYL has been hounding you about.

Materials

Copper and brass is used exclusively in this antenna. These metals get along together, so dissimilar metal corrosion is eliminated. Both metals solder well, too. **Table 1** provides a detailed parts listing for the antenna.

Construction

Cut the copper tubing to the lengths indicated in Table 1. Item 9 is a $1\frac{1}{4}$-inch nipple cut from the 20 inch length of $\frac{1}{2}$-inch tubing. This leaves $18\frac{3}{4}$ inches for the $\lambda/4$-matching stub. Item 10 is a $3\frac{1}{4}$-inch long nipple cut from the 60-inch length of $\frac{3}{4}$-inch tubing. The $\frac{3}{4}$-wave element should measure $56\frac{3}{4}$ inches long. Remove burrs from the ends of the tubing after cutting, and clean the mating surfaces with sandpaper, steel wool, or emery cloth.

After cleaning, apply a very thin coat of flux to the mating elements and assemble the tubing, elbow, tee, endcaps, and stubs. Solder the assembled parts with a propane torch and rosin-core solder. Wipe off excess solder with a damp cloth, being careful not to burn yourself. The copper tubing will hold heat for a long time after you've finished soldering. After soldering, set the assembly aside to cool.

Flatten one each of the $\frac{1}{2}$-inch and $\frac{3}{4}$-inch pipe clamps. Drill a hole in the flattened clamp as shown in Fig 1B. Assemble the clamps and cut off the excess metal from the flattened clamp using the unmodified clamp as a template. Disassemble the clamps.

Assemble the $\frac{1}{2}$-inch clamp around the $\frac{1}{4}$-wave element and secure with two of the screws, washers, and nuts as shown in Fig 1B. Do the same with the $\frac{3}{4}$-inch clamp around the $\frac{3}{4}$-wave element. Set the clamps initially to a spot 4 inches or so above the bottom of the "J" on their respective elements. Tighten the clamps only finger tight, since you'll have to move them when tuning.

Tuning

Tuning an antenna couldn't be simpler.[2] The toughest part might be determining what type feed line you are going to use. Anything from RG-58 to open-wire line is usable. The J-Pole can be fed directly from 50 Ω coax alone or with a $^1/_2$-wave balun, or twin lead, or whatever. Before tuning, mount the antenna vertically, about 5 to 10 feet from the ground. A short TV mast on a tripod works well for this purpose.

When tuning VHF antennas, keep in mind that they are extremely sensitive to nearby objects—such as your body. Attach the feed line to the clamps on the antenna, and make sure all the nuts and screws are at least finger tight. If using coax, it really doesn't matter to which element ($^3/_4$-wave element or stub) you attach the coaxial center lead. I've done it both ways with no variation in operational effectiveness. Tune as follows:

1. Apply RF at the frequency you want the antenna to perform best.
2. Check the SWR.
3. Turn off the RF.
4. Move the clamps *equally* up from the original position $^1/_2$ inch.
5. Reapply RF.
6. Check the SWR again, and turn off the RF.
   a. If the SWR went higher, move the clamps 1 inch downward, equally.
   b. If the SWR went lower, move the clamps $^1/_2$ inch upward, equally.
7. Reapply RF.
8. Check the SWR once more, then turn off the RF. By this time you will be approaching minimum SWR (You may *never* get a 1:1 SWR. Don't sweat it, it's not important enough to worry about.)
9. Adjust the clamps a small amount in the direction of minimum SWR.
10. Repeat Steps 7 through 9 until minimum SWR is achieved.
11. Remove RF.

A point to remember about tuning the antenna is this: It will tune with any type of feed line, but the clamps will not be in the

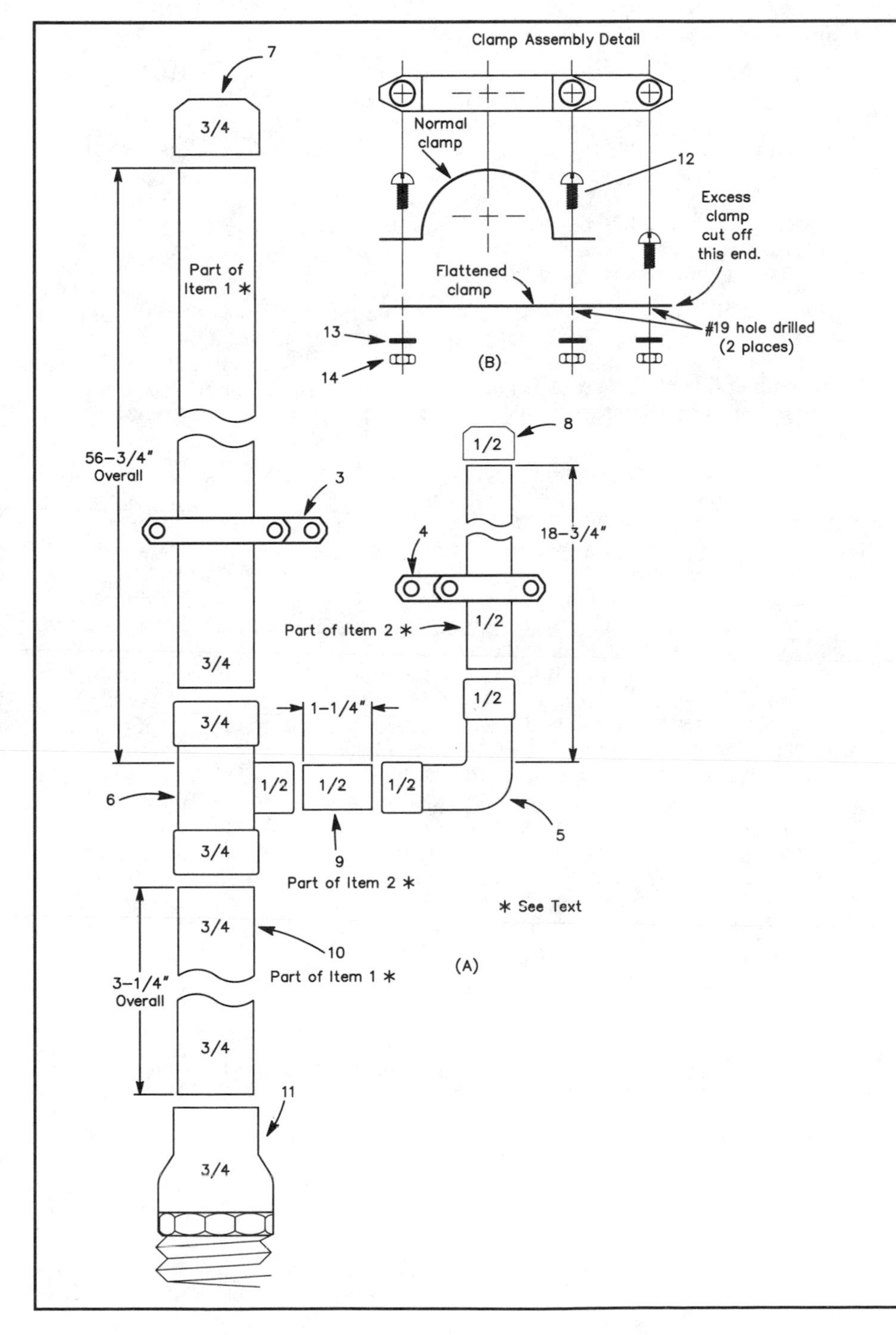

Fig 1—(A) Exploded assembly diagram of all-copper J-Pole antenna. Item numbers refer to parts list in Table 1. (B) Detail of clamp assemblies. Both clamp assemblies are the same.

Item	*Qty*	*Part or Material Name*
1	1	$^3/_4$ inch × 10 ft length of rigid copper tubing (enough for 2 antennas, 60 inches per antenna)
2	1	$^1/_2$ inch × 10 ft length of rigid copper tubing (enough for 6 antennas, 20 inches per antenna)
3	2	$^3/_4$ inch copper pipe clamps
4	2	$^1/_2$ inch copper pipe clamps
5	1	$^1/_2$ inch copper elbow
6	1	$^3/_4$ × $^1/_2$ inch copper tee
7	1	$^3/_4$ inch copper end cap
8	1	$^1/_2$ inch copper end cap
9	1	$^1/_2$ × 1$^1/_4$ inch copper nipple (Make from item 2. See text)
10	1	$^3/_4$ × 3$^1/_4$ inch copper nipple (Make from item 1. See text)
11	1	Your choice of coupling to mast fitting ($^3/_4$ × 1 inch NPT used at KD8JB)
12	6	# 8-32 × $^1/_2$ inch brass machine screws (round, pan, or binder head)
13	6	# 8 brass flat washers
14	6	# 8-32 brass hex nuts

same place for all types of line. In other words, 50-Ω coaxial cable will tune further down the antenna than 600-Ω open-wire line. You have to think of this feedpoint as being similar to a delta match, which it is, except the elements run parallel and not away from each other.

Final Assembly

The final assembly of the antenna will determine its long-term survivability. Perform the following steps with care:

1. After adjusting the clamps for minimum SWR, mark the clamp positions with a pencil.
2. Remove the feed line and clamps.
3. Apply a very thin coating of flux to the inside of the clamp and the corresponding surface of the antenna element where the clamp attaches.
4. Install the clamps and tighten the clamp screws. Don't attach the feed line yet, and leave off the final washer and nut at the feedline attachment point.
5. Solder the clamps where they are attached to the antenna elements.
6. Apply a small amount of solder around the screw heads and nuts where they contact the clamps. Don't get solder on the screw threads!
7. Clean away excess flux with a non-corrosive solvent.

After final assembly and erecting/mounting the antenna in the desired location , attach the feed line you tuned the antenna for, and secure with the remaining washer and nut. It would be a good idea to weather-seal this joint with RTV. Otherwise, you may find yourself repairing the feed line after a couple years.

On-Air Performance

Years ago, prior to building the first J-Pole antenna for this station, I used a standard 1/4-wave ground plane vertical antenna. While there is no problem working the various repeaters around town with my 1/4 wave antenna, simplex operation left a lot to be desired, so I felt something with a little more gain was necessary. Hence the switch to the J-Pole. In on-air comparisons with the Ringo Ranger 2B, a popular antenna everywhere, a small difference in bandwidth was noted. This J-Pole's bandwidth, as built here, is slightly wider, probably from the greater element thickness, resulting in a lower Q. Actual performance differences between antennas of similar dimensions, such as those of the Ringo Ranger are negligible, although significantly better than the 1/4-wave ground-plane vertical.

Post Inspection/Evaluation of the Original J-Pole

1. ALUMINUM MOUNTING PLATE: The plate was rust-stained from contact with the non-stainless steel elements of the antenna. The alloy used for the plate was 2024 aluminum, which is quite hard and holds its finish well. However, the rust deposits affect the continuity between elements. This plate was reused after cleaning.
2. GALVANIZED PAINTED 3/4-WAVE SECTION: The painted portion of this element survived the ravages of weather very well. The small portion where the paint was removed to make contact with the aluminum plate was completely coated with rust, as were the U-bolts holding the element to the plate, and the screw (probably zinc plated) used to ensure contact between the section and the mounting plate's upper edge. The hole drilled for the stainless screw used to attach the coaxial cable shield was in as good a condition as when originally drilled owing to the liberal application of Silastic 732 RTV[4] after assembly. This section was reused as a 5-foot mast section. I would use 1 1/4-inch aluminum tubing if this antenna were rebuilt in this form.
3. NON-STAINLESS HARDWARE: All plated and unplated hardware exhibited varying degrees of oxidation (rust). Some of the galvanized hardware came through with only spot rust where the galvanized coating had either deteriorated or been scraped away. The cadmium-plated mast brackets bought at Radio Shack showed rust only where excess length on the U-bolts was removed and at the bends in the bracket plates. These brackets were reusable, but the nuts were replaced. The U-bolts holding the 3/4-wave section to the mounting plate were so badly rusted that when removal was attempted, the bolts broke off.
4. STAINLESS STEEL HARDWARE: As expected, all stainless steel came through in great shape. Even though mated with some non-stainless hardware like star washers or nuts with integral star washers, the stainless hardware was easy to disassemble. Never use other than stainless steel or brass hardware in antenna construction! It's well worth the extra cost.
5. COAXIAL CABLE: This antenna was fed with RG-8X (foam) made for LaCue Communications. Some cable was cut off for inspection when the antenna was taken down. (I do this whenever I take down an antenna to get a visual idea of how the coax is doing.) The cable was terminated at the antenna with crimp lugs, which I crimped and soldered. The shield remained in good condition throughout except for normal copper oxide near the lug. (I used RTV[4] to seal the spot where the center lead passed through the shield. The sealant worked well.) However, the foam dielectric between the shield and the terminal lug cracked in a number of places causing water to leak in and wick down the center conductor. Not good! Interestingly, the foam that had melted near the soldered terminal lug protected the center conductor well at that point.
   Lesson: When terminating coax in this fashion, the foam dielectric should be sleeved or taped to prevent cracking. In your haste to use a new antenna, always solder the lugs—never leave them just crimped. I should point out that I was responsible for the deterioration of the coaxial cable because of the way I used it. LaCue had nothing to do with what happened in this instance. I use LaCue's RG-8X cable for most of my feed lines, and when properly terminated, it works very well.

References

[1]M. P. Hood, "All-Metal 2-Meter J-Pole Antenna" (and References), *Ham Radio*, Jul 1984, pp 42-44.

[2]"A Combination 6 and 2 Meter J-Pole," *FM and Repeaters* (Newington: ARRL, 1972). (Out of print.)

[3]"Building and Using VHF Antennas," *VHF Handbook* (Newington: ARRL, 1972). (Out of print.)

[4]Silastic 732 RTV is made by the Dow Corning Company.

The Double Cross Vertical Antenna

By Robert Wilson, AL7KK
Box 110955
Anchorage, AK 99511

Antennas have been part of my business for years and I have placed my designs at large stations in a number of countries. (Money is no object with the big commercial arrays.) Even though I design antennas for a living, I still like to play with them at home. Part of the fun of antenna construction is producing a high quality, efficient antenna that is truly low in cost.

One snowy Alaskan afternoon I was inspired to design a home-brew antenna project. Simplicity was the primary consideration. It had to be something I could build with just a couple of coax connectors, a few coat hangers and my trusty soldering iron. It also had to be theoretically feasible and efficient. Despite my best intentions, my antennas have a tendency to grow like rabbits. Within a short period of time my living room was filled with paper and wire!

The Double Cross Design

Where would antenna designers be without computers? After a bit of work at my keyboard I soon developed the "X" antenna. As I examined the design I noticed a unique property of the "X": It could be mounted on a metal pole with almost no interaction. (The vertical center line between elements was also a null line.) Stacking the antenna seemed natural, and after looking at the result (see Fig 1) the name "Double Cross" seemed natural too!

This antenna is a variation on a vertical dipole. Imagine two V-shaped wires, one opening upward and one downward. Now imagine that both V wires are connected to a coax cable at their apexes. The coax shield is soldered to one V and the coax center wire to the other V.

By adjusting the angle of the V to 70°, the antenna can be made to resonate at the desired frequency and that angle will give the best bandwidth. The angle does not have to be extremely precise. Angles from about 60 to 110 degrees do little to change the feed impedance. (The impedance remains about 30 to 35 Ω over these angles.) The mid-line between the V wires is a null line. This permits the support pole and coax feeder harness to be mounted in the center line area with no problems.

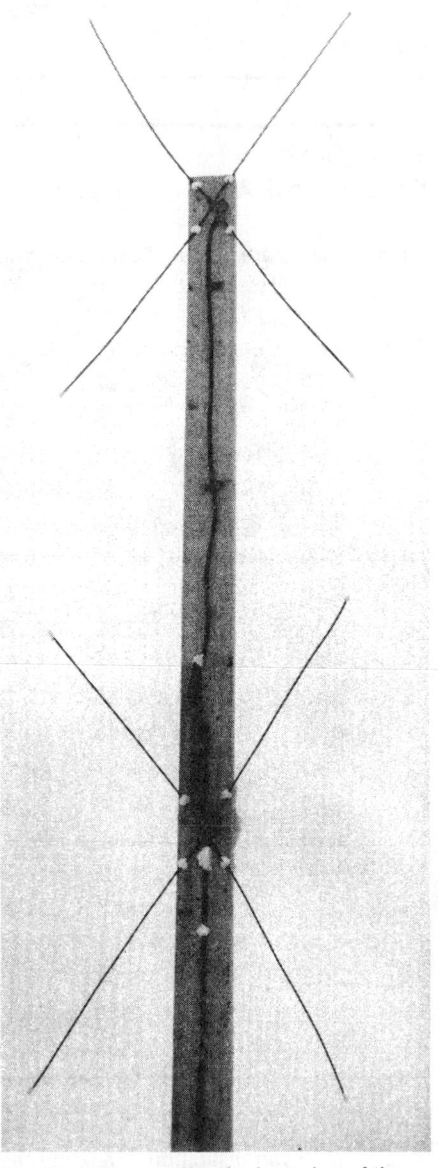

The 2-meter experimental version of the Double Cross antenna made with no. 10 wire, coax, silicone glue and a 2 × 4 board.

The angle of the V apex is set easily by adjusting the tip-to-tip spacing of the V elements. On paper and in actual practice it seems to be reasonably noncritical. For example, a tip-to-tip error of ±5% seems to make little difference in the operation of the antenna.

It also occurred to me that grounding the topmost elements on the antenna would offer lightning protection for the receiver. Turning the top dipole upside down and mounting the hot element of the next lower dipole *upward* also seemed to be a neat symmetrical method to balance the antenna currents. Feeding the dipoles 180° out of phase is all that is required for both protection and balance. This can be easily accomplished by adjusting the length of the feed harness (see Fig 2).

Old hams told me that using two wires in a V would not give a circular pattern because it was not balanced. This "old ham's tale" is simply not true; the pattern of the Double Cross is only 0.5 dB out of round. For the sake of a little irregularity there isn't a compelling reason to increase the complexity by adding more wires. To prove that more complex designs were not necessary, I constructed a three-dimensional version of the antenna and found it was indeed very difficult to handle. After a thorough examination of the calculations and construction of the flat "X" antenna, I proved that a two-dimensional design was entirely sufficient.

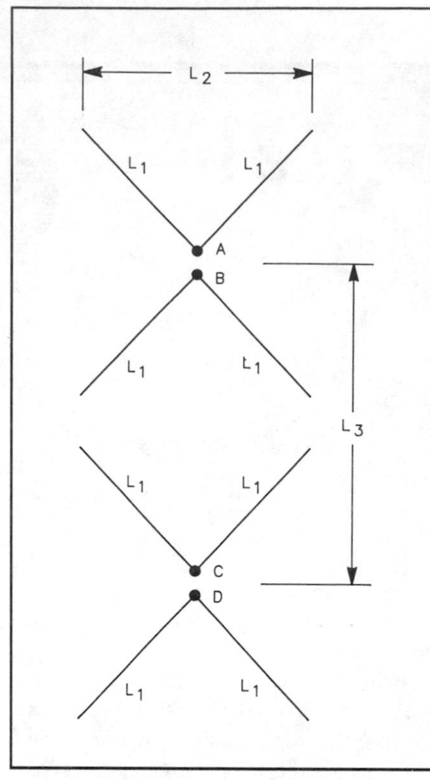

Fig 1—Double Cross antenna design.

After I built the first coat-hanger "X" antenna I also discovered that the measured bandwidth was better than expected. The element diameter was increased to ½ inch on the 2-meter theoretical model and this allowed the antenna to cover more than the full band.

Table 1 provides element lengths, tip-to-tip spacings, element diameters and phasing-line information for most ham bands. Table 2 offers the same information for 1 to 30 MHz in arbitrarily numbered bands. Incidentally, using wire for the elements is quite acceptable even though tubing is indicated. For ham operation a sufficient bandwidth can be obtained by using thin wire elements.

A single stacked "X" dipole could be strung between two trees for low-frequency operation. Attempting the Double Cross stacking arrangement on 80 or 160 meters would be unreasonable, but a single unstacked "X" would perform as an excellent wide-band vertical on these bands.

It may be a good idea to employ a matching transformer between the 50-Ω coax and the antenna. Such an RF transformer can be made by using a high quality, high frequency powdered core with a cross section of at least ½ by ¼ inch for 150 watts. For the HF bands the 50-Ω primary should be 10 turns and the 30-Ω secondary should be 8 turns. I prefer to use 18 gauge wire with Teflon insulation, wrapping the first winding and then interlacing the second. I also like to tie down the ends with fishing line and coat the transformer with a heavy layer of clear silicone glue. A properly constructed transformer should last up to 50 years—if it doesn't take a direct lightning strike!

Table 1
Double Cross Antenna Lengths for Amateur Bands

All dimensions are in feet. "Bandwidth" shows the lower and upper 2:1-SWR frequencies, MHz.

Band	Freq.	*L1* *Element*	*L2* *Tip-tip*	*L3* *Spacing*	*L4* *Coax 1*	*L5* *Coax 2*	*L6* *Diam.*	*Bandwidth*
160	1.90	111.05	127.38	323.68	256.36	85.45	3.04	1.7—2.1
80	3.70	57.03	65.41	166.22	131.64	43.88	1.56	3.3—4.1
75	3.90	54.10	62.06	157.69	124.89	41.63	1.48	3.5—4.3
40	7.15	29.51	33.85	86.01	68.12	22.71	0.81	6.4—7.9
30	10.13	20.84	23.90	60.74	48.11	16.04	0.57	9.1—11.1
20	14.18	14.89	17.07	43.39	34.36	11.45	0.41	12.8—15.6
17	18.11	11.65	13.36	33.96	26.90	8.97	0.32	16.3—19.9
15	21.23	9.94	11.40	28.98	22.95	7.65	0.27	19.1—23.3
12	24.93	8.46	9.71	24.67	19.54	6.51	0.23	22.4—27.4
10	28.50	7.40	8.49	21.58	17.09	5.70	0.20	25.7—31.4
6	50.10	4.21	4.83	12.28	9.72	3.24	0.12	45.1—55.1
2	146.00	1.45	1.66	4.21	3.34	1.11	0.04	131.4—160.6

Table 2
Double Cross Antenna Lengths for Various SWL Bands

All dimensions are in feet. "Bandwidth" shows the lower and upper 2:1-SWR frequencies, MHz

Band	Freq.	*L1* *Element*	*L2* *Tip-tip*	*L3* *Spacing*	*L4* *Coax 1*	*L5* *Coax 2*	*L6* *Diam.*	*Bandwidth*
1	1.00	211.00	242.02	615.00	487.08	162.36	5.77	0.9—1.1
2	1.20	175.83	201.68	512.50	405.90	135.30	4.81	1.1—1.3
3	1.44	146.53	168.07	427.08	338.25	112.75	4.01	1.3—1.6
4	1.73	121.97	139.89	355.49	281.55	93.85	3.34	1.6—1.9
5	2.07	101.93	116.92	297.10	235.30	78.43	2.79	1.9—2.3
6	2.49	84.74	97.20	246.99	195.61	65.20	2.32	2.2—2.7
7	2.99	70.57	80.94	205.69	162.90	54.30	1.93	2.7—3.3
8	3.58	58.94	67.60	171.79	136.06	45.35	1.61	3.2—3.9
9	4.30	49.07	56.28	143.02	113.27	37.76	1.34	3.9—4.7
10	5.16	40.89	46.90	119.19	94.40	31.47	1.12	4.6—5.7
11	6.19	34.09	39.10	99.35	78.69	26.23	0.93	5.6—6.8
12	7.43	28.40	32.57	82.77	65.56	21.85	0.78	6.7—8.2
13	8.92	23.65	27.13	68.95	54.61	18.20	0.65	8.0—9.8
14	10.70	19.72	22.62	57.48	45.52	15.17	0.54	9.6—11.8
15	12.84	16.43	18.85	47.90	37.93	12.64	0.45	11.6—14.1
16	15.41	13.69	15.71	39.91	31.61	10.54	0.37	13.9—17.0
17	18.49	11.41	13.09	33.26	26.34	8.78	0.31	16.6—20.3
18	22.19	9.51	10.91	27.72	21.95	7.32	0.26	20.0—24.4
19	26.62	7.93	9.09	23.10	18.30	6.10	0.22	24.0—29.3
20	31.95	6.60	7.57	19.25	15.25	5.08	0.18	28.8—35.1

Design Calculations

A simple four-function calculator can be used to calculate the antenna dimensions.

1) The lengths of each leg of the V elements (L1):

$$L1 = \frac{64.2 \text{ meters}}{f_{MHz}} = \frac{211 \text{ feet}}{f_{MHz}}$$

2) The tip-to-tip distance (L2) of the open end of the V element gives a 70° apex angle:

L2 = 1.147 × L1 (results in either feet or meters, according to the original L1 values)

3) The separation of two "X" elements is ⅝ λ or L3:

$$L3 = \frac{187.0 \text{ meters}}{f_{MHz}} = \frac{615 \text{ feet}}{f_{MHz}}$$

4) L4 represents the length of the 50-Ω solid polyethylene dielectric coax (velocity factor 0.66) from the **upper** dipole to the summing junction where upper and lower dipoles are connected:

$$L4 = \frac{224.4 \text{ meters} \times 0.66}{f_{MHz}} = \frac{738 \text{ feet} \times 0.66}{f_{MHz}}$$

5) L5 represents the length of 50-Ω solid polyethylene dielectric coax from the **lower** dipole to the summing junction mentioned above, *and* the length of the two parallel 70-Ω coaxial cables used for impedance matching:

$$L5 = \frac{75 \text{ meters} \times 0.66}{f_{MHz}} = \frac{246 \text{ feet} \times 0.66}{f_{MHz}}$$

6) L6 represents the diameter of the V legs required for the indicated bandwidth. However, ordinary wire also works well for practical ham antennas:

$$L6 = \frac{1.759 \text{ meters}}{f_{MHz}} = \frac{69.2 \text{ in.}}{f_{MHz}} = \frac{5.77 \text{ ft}}{f_{MHz}}$$

Double Cross Construction

Construction of the first "X" antenna was accomplished with coat-hanger wire soldered to a coax connector. The coax line was connected and the free end was pulled through a ½-inch support pipe for testing. The results were excellent. The SWR was low and the bandwidth was exactly as calculated.

The next step was construction of a stacked 2-meter Double Cross antenna with a complete coaxial feed harness. This was done using no. 10 Copperweld wire salvaged from an open-wire telephone system. It was screwed to a 2 × 4 board and glued in place with silicone adhesive. There is no question that this was a minimum cost antenna! Once again, everything worked fine and the SWR was 1.1:1 over the entire 2-meter band.

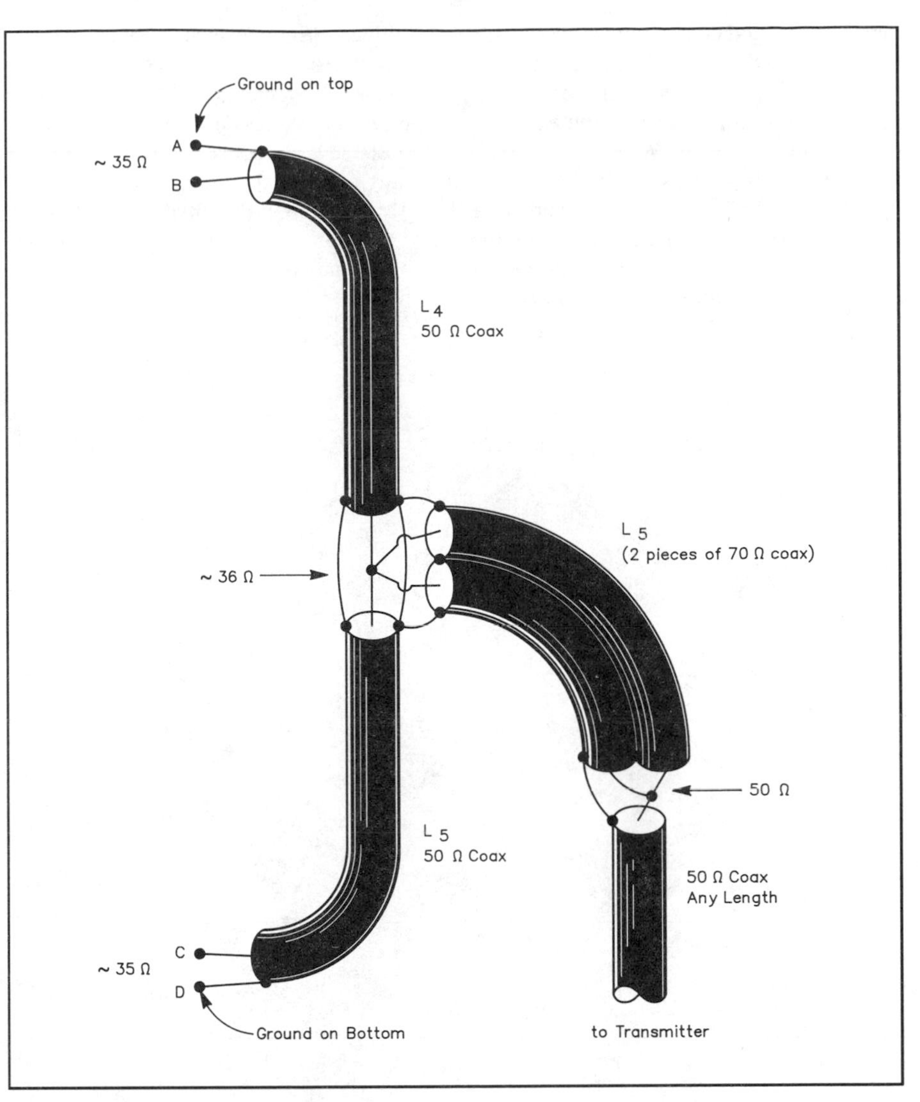

Fig 2—Double Cross coax harness.

The stacking harness shown in Fig 2 is designed for a 180° phase reversal. The shield of the ¾-λ section of 50-Ω coax goes to the top V. This is length L4. The shield of the ¼-λ 50-Ω coax section goes to the bottom V. This is length L5.

The two 50-Ω phasing lines are soldered together at a common point. The impedance at this point becomes about 36 Ω because there are two 50-Ω lines in parallel.

Matching is easy with a "Q" section, a ¼-λ series matching section made from 42-Ω coax. Two pieces of solid polyethylene coax are cut to length L5 and soldered in parallel (shield to shield and center to center) to make the "Q" section. I used two pieces of 70-Ω coax and achieved an excellent match, but it is possible to substitute one length of 95-Ω coax if you need to improve the match further. Examples of 95-Ω coax are RG-180B or RG-195A. Alternatively, it is possible to build short pieces of 42-Ω coax from brass hobby tubing. The inner conductor needs to be 0.217 inch or about 7/32 inch, and the outer conductor's inner diameter should be 0.5 inch. However, I recommend starting with two parallel 70-Ω coax cables first.

All harness connections should be kept short. I prefer to connect all center conductors first. Make one last check against Fig 2 to be sure that everything is correctly connected. Then smooth the joints and wrap them tightly with Teflon plumber's tape. Now connect the shields, taking care not to melt the polyethylene insulation. A wet cloth will quickly cool the joint after soldering. Let it cool for several minutes because the polyethylene core cools much more slowly than copper. The joint can then be dried and coated with silicone glue to make it waterproof. If you insist on a neater appearance, slide some shrink tubing over the joint before the silicone hardens and shrink it. Wipe off the extra silicone for a first-class job.

The final step is to use a VOM and make a resistance check of the antenna with the coax in place. The path from the center conductor of the coax to the shield should exhibit an infinite resistance (open circuit). The path from the center conductor to the two inside V elements should show a short. The resistance from the coaxial shield to the outside V elements should also indicate a short.

Route all coax straight down the middle line of the antenna. Secure it in place so that wind, ice and time will not change its location.

These Double Cross stacked dipoles have given me the extra low-angle gain necessary for improved 2-meter coverage. I keep thinking about how well a long dipole stack would perform on the UHF bands and how nicely a 20-meter Double Cross could work—if I could only get two tall trees to grow in my muskeg swamp!

A Vertical 6-Meter Wire Extended Double Zepp

By Wayde S. Bartholomew, WA3WMG
RD3 Box 3769
Pottsville, PA 17901

WA3WMG describes his simple, low-cost but very effective 6-meter wire antenna for repeater coverage.

I needed a low-cost gain antenna with vertical polarization for 52.525 MHz. The concept of an Extended Double Zepp looked interesting. I constructed a mock-up antenna at my QTH to compare it with a quarter-wave vertical at a height of 50 feet. The Extended-Double Zepp typically outperformed the vertical by two to three S units for both receive and transmit. This was with a power output of 25 W, talking with stations 30 to 100 miles distant. This encouraged me to construct the final version, which was placed at the 100-foot level on a commercial tower 1800 feet above sea level.

Construction is fairly straightforward. See **Fig 1**. The upper and lower legs of the dipole sections are supported by 1½-inch PVC pipe. The 1:1 balun and matching section are placed inside another piece of 1½-inch PVC pipe for support and weather protection.

Spacing from the tower has a definite effect on directivity. **Fig 2** shows the effect on the azimuthal and elevation patterns of spacing the antenna 5 feet away from a tower, compared to a 10-foot spacing from the tower. In this case the bottom of the antenna is at the 50-foot level on a 100 foot high tower. The front-to-back ratio is about 9 dB for the 5-foot spacing.

Although it is still distorted, the pattern for the 10-foot spacing is more omnidirectional in nature than the 5-foot spacing. Of course, the larger spacing does present more of a construction problem when using PVC pipe.

Tuning of the antenna is simple. The legs are temporarily stretched out on the tower with the bottom of the antenna a few feet off the ground. The ladder line is purposely made a little long at installation (3 feet is a good starting point), and then trimmed a little at a time for the lowest SWR. This should be close to 1:1.

Once the feed line is trimmed, the ends of the PVC pipe should be sealed with RTV sealant. A small hole should be drilled in the bottom of the PVC to drain off any condensation that may form inside. A small screen should be placed over the drain hole to keep out spiders and wasps—they seem to like antennas for homes!

Results have been impressive. During band openings the antenna is an excellent performer, and during marginal band conditions it has the gain to still be effective. In three months time from November 1991 to January 1992, 25 states were worked. Three other Extended Double Zepps were constructed for repeater use, with excellent results.

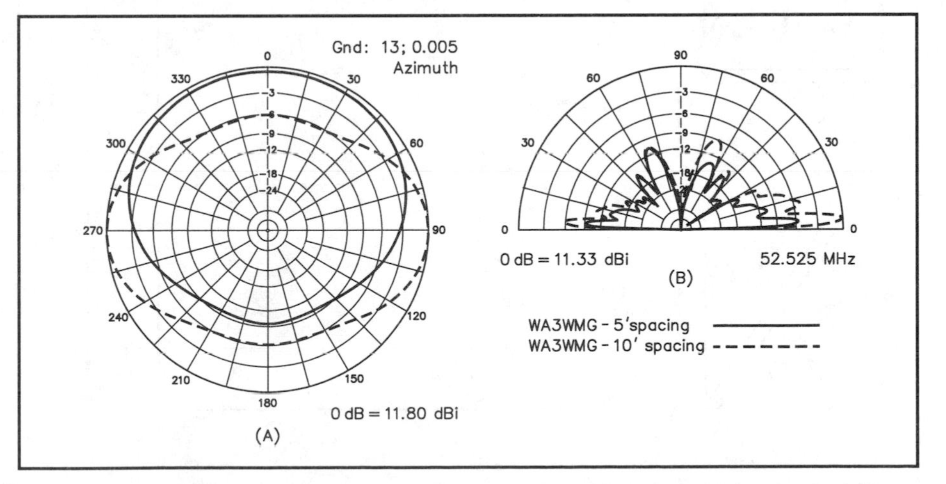

Fig 2—Patterns for vertical 6-meter EDZ mounted parallel to metal tower. At A is a comparison of computed azimuthal patterns for 5- and 10-foot spacings from tower; at B is a comparison of elevation patterns. The larger spacing yields coverage that is more omnidirectional.

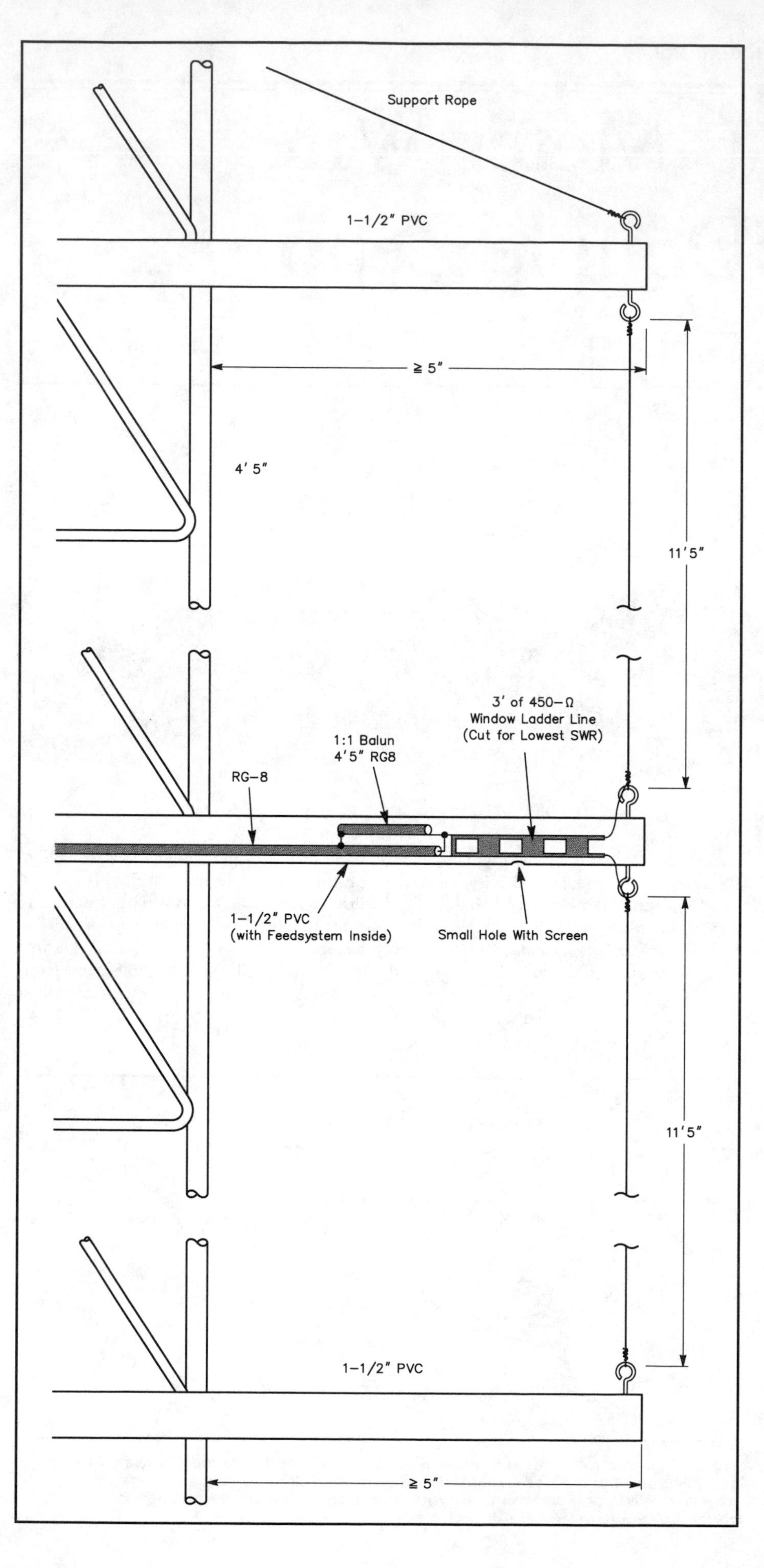

Fig 1—Mounting dimensions for 6-meter Extended Double Zepp antenna. The 450-Ω "window" ladder line is cut long initially at 3 feet and trimmed for best SWR at installation. The feed system, including the 1:1 coax balun and window line, are placed inside 1½-inch PVC pipe for support and weather protection.

Chapter 3

HF

from September 1981 *QST*

A Modest 45-Foot DX Vertical for 160, 80, 40 and 30 Meters

If it's DX you want, this low-angle radiator will put it in your lap! Build it now and collect DX dividends this winter.

By Wayne H. Sandford, Jr.,* K3EQ

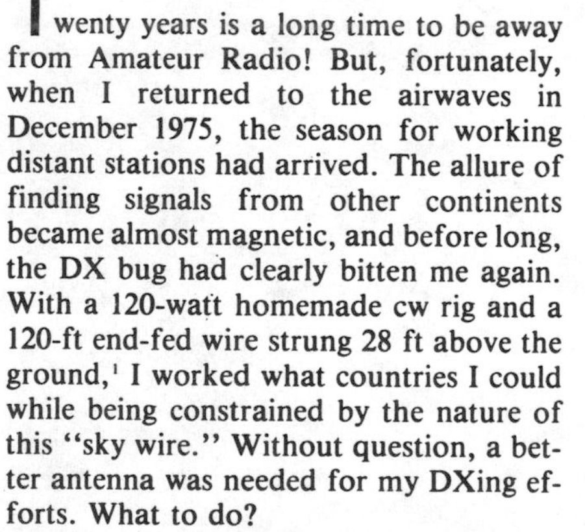

Twenty years is a long time to be away from Amateur Radio! But, fortunately, when I returned to the airwaves in December 1975, the season for working distant stations had arrived. The allure of finding signals from other continents became almost magnetic, and before long, the DX bug had clearly bitten me again. With a 120-watt homemade cw rig and a 120-ft end-fed wire strung 28 ft above the ground,[1] I worked what countries I could while being constrained by the nature of this "sky wire." Without question, a better antenna was needed for my DXing efforts. What to do?

Improvements began with the construction of a 36-ft wooden tower I built to support a 2-element quad for 10, 15 and 20 meters. From the top of this tower, I hung a 40-meter vertical antenna, followed by the installation of twenty-four 50-ft radials. DXing on 40 meters improved noticeably as a result of this effort.

* 59 S Woodridge Dr, Warrington, PA, 18976

For awhile I was satisfied to leave my 80-meter inverted L alone. It was strung between the quad tower and a mast supporting one end of my end-fed wire. Admittedly, results with this antenna were mediocre. During the winter of 1979-80, as I approached the requirements for Five-Band DXCC on all bands except 80 meters (only 50 confirmed), I began to think about better DX antennas for the lower frequencies.

Research

I looked through back issues of *QST* and other publications for antenna articles: A *QST* article by Hollander[2] triggered thoughts of constructing a multiband vertical antenna. Radiation patterns of 1/8-, 1/4-, 1/2- and 5/8-wavelength vertical antennas indicate that an antenna having this configuration would give low-angle radiation on four bands. Calculations indicated these fractional lengths could be applied to 160-, 80-, 40- and the new 30-meter band that will become available sometime during 1982. A 5/8-wavelength vertical antenna for 30 meters is nearly 60 ft high. A half wavelength for 40 meters is 70 ft; 1/4 wavelength for 80 meters is 70 ft; and 1/8 wavelength on 160 meters is 68 ft. Therefore, if a pole 60 ft high were used, series inductance could be added to obtain the required electrical length on all four bands. But as much as I desired to have a vertical antenna 60 ft tall or greater, I decided to see if an antenna as short as 40 ft would serve my purpose. Furthermore, although not too much has been said by the neighbors about the 2-element quad, I feared that a 60-ft vertical antenna might stimulate a barrage of adverse comments!

After pondering the matter for some time and studying radiation resistance and reactance plots for vertical antennas,[3] the solution of the problem came into focus. For an antenna shorter than 60 ft some form of loading was needed. A "top hat" provides an efficient means for doing this.[4]

This multiband antenna should first be calculated for 5/8 wavelength on 30 meters. It will give an almost perfect match to a 50-ohm line by adding a small

Table 1

Dimensions for Optimum Height of the Vertical Radiator

Radiator Height (ft)	*Top-hat dia (ft)*	*Calculated Heights for 10.125 MHz (Deg.) (Sum = 225 Deg.)*		*Calculated Heights for 7.025 MHz (Deg.)*			*Calculated Heights for 3.525 MHz (Deg.)*			*Calculated Heights for 1.8125 MHz (Deg.)*		
		Height	Top Loading	Height +	Top Loading	= Sum	Height +	Top Loading	= Sum	Height +	Top Loading	= Sum
43	11	159.5	65.5	110.6	56.7	167.3	55.5	37.4	92.9	28.5	21.4	49.9
44	9.6	163	62	113.2	52.6	165.8	56.8	33.2	90	29.2	18.6	47.8
45	8.2	166.9	58.1	115.8	48.1	163.9	58.1	29.2	87.3	29.9	16	45.9

Meters = feet × 0.3048

Table 2

How Top-Hat Loading is More Effective on Lower Bands in Increasing Effective Height

F (MHz)	*Top Loading (Deg.)*	*Top Loading (ft)*	*Ant. Effect. Height (ft)*	*Ant. Effect. Height (λ)*
1.8125	18.6	28	72	0.133
3.525	33.2	25.7	69.7	0.249
7.025	52.6	20.5	64.5	0.456
10.125	62	16.7	60.7	0.625

Meters = feet × 0.3048

inductance in series with the antenna at the feed point, then tuning out the capacitive reactance with a shunt inductor. If the antenna is a half-wavelength long at 40 meters (the length at which reactance is zero), it could be adjusted easily by using a parallel-tuned tank in series with the ground lead, and by tapping the feed line at a point on the tank just a few turns up from the ground end. The tap and tuning adjustments are arranged to give the best match. It seemed that if the antenna were 1/4 wavelength long at 80 meters, it could be increased in length to provide a 50-ohm feed point by means of a small series inductor and a shunt capacitor to tune out the reactance. In addition, since it would be considerably shorter than 1/4 wavelength on 160 meters (on the order of 1/8 wavelength) it could be made to look like a 1/4-wavelength antenna by adding series inductance to ground. Matching could be effected by tapping the line a few turns up on the coil. Many dyed-in-the-wool DXers would not consider a 1/8-wavelength vertical antenna, but Sevick[5] has shown that this can be an efficient radiator when used with an effective ground system and a low-loss, base-loading inductor.

Design Procedure

I could not remember having seen details of vertical antennas that explained how to calculate the effect of the "top hat." But in past issues of *QST* I found an article by Schulz,[6] which was just what I needed. Although his design was for a 1/4-wavelength antenna, the equations are presumed applicable for calculating the "top-hat" effects on 1/2- and 5/8-wavelength antennas. Calculations with his equations indicated that a 44-ft vertical antenna loaded by a 9.6-ft diameter "top hat" would give the results I wanted. My aim was to have a vertical antenna that would be 5/8 wavelength on 30 meters, 1/2 wavelength on 40 meters, 1/4 wavelength on 80 meters and 1/8 wavelength on 160 meters. Table 1 shows calculated electrical lengths and required "top-hat" diameters for vertical radiators from 43 to 45 ft high, showing that the 44-ft height is about right to give the required four-band performance. Table 2 shows that the "top hat" is more effective in increasing the length of the radiator as the frequency goes down.

Since this design promised a high degree of success, the preliminary circuit (Fig. 1) was prepared. A parts list was compiled (Table 3), and material collection was begun.

Construction

Purchases for the project included a 40-ft telescoping TV mast (its extended length turned out to be 38.5 ft) and a 6-ft galvanized fence post, which would just fit inside the lower mast section. With 6 in. of the post telescoped inside the mast,[7] the overall length was the required 44 ft. To secure the mast to the fence post, two slits were made in the lower section of the mast with the aid of a hacksaw. A stainless-steel radiator hose clamp and a 1/4-20 bolt, 2-1/2 in. long, were used to clamp the mast firmly to the fence post.

The eight-spoke "top-hat" is constructed in a manner similar to that used by Hollander.[8] There are eight 5-ft lengths of 1/2-in. diameter conduit fastened to an 11-in. square, 1/8-in. thick aluminum plate. The spokes are held firmly against the plate by means of 6-32 stainless-steel hardware. Aluminum angle stock is used to fasten the plate to the top of the upper mast section. This stock, which is 1/8 in. thick by 1-1/2 in. wide, is cut into four 1-in. lengths. Two 1/4-20 stainless-steel bolts, 2 in. long, are used to fasten the angles to the upper mast section. I suggest the use of lock washers in all cases where the bolts are used. Good electrical contact can be assured by connecting all "top-hat" radials together and to the mast with 1/4-in. wide braid using stainless-steel, self-tapping screws. Three 48-in. long heavy-duty, screw-in steel anchors are used for the guy points. They are located 25 ft from the tower base. Four sets of guys are used. They are made from no. 12-1/2-gauge steel wire. A total of 42 egg insulators are installed to break the guys into lengths no longer than 19 ft. The base of the mast sits on a 7-in high, heavy-duty standoff insulator, which in turn rests on a 6-in. diameter concrete base that is 3 ft deep, with 4 in. protruding above ground.

Installation of the Mast

First, stand the mast upright and attach the lower set of guys to the anchors. The three top sections of the mast are pushed up from a ladder resting against the mast. Proceed by attaching the next set of guy wires to the anchors. The ladder is then extended to the second guy level, and the upper section is pushed up next. A piece of 1/4-in. braid is fastened across the joint between the top and second section of the mast, using self-tapping, stainless-steel screws. Follow this by pushing the second and top sections up together. Next, a strap is connected across the other two joints to ensure good electrical contact. Complete this part of the installation by connecting all guys to the mast, then adjust them so that the mast stands vertically.

All tuning components are mounted in a fiberglass box. Fig. 2 shows the open tuning box and components. Fig. 3, a photograph of the base of the antenna, shows how the box is attached to the 3/4-in. galvanized water pipe ground rod, and how the radial wires are terminated on a square aluminum plate (similar to the method used by Sevick).[9] The plate is fastened to the ground rod with aluminum angle brackets, stainless-steel hardware

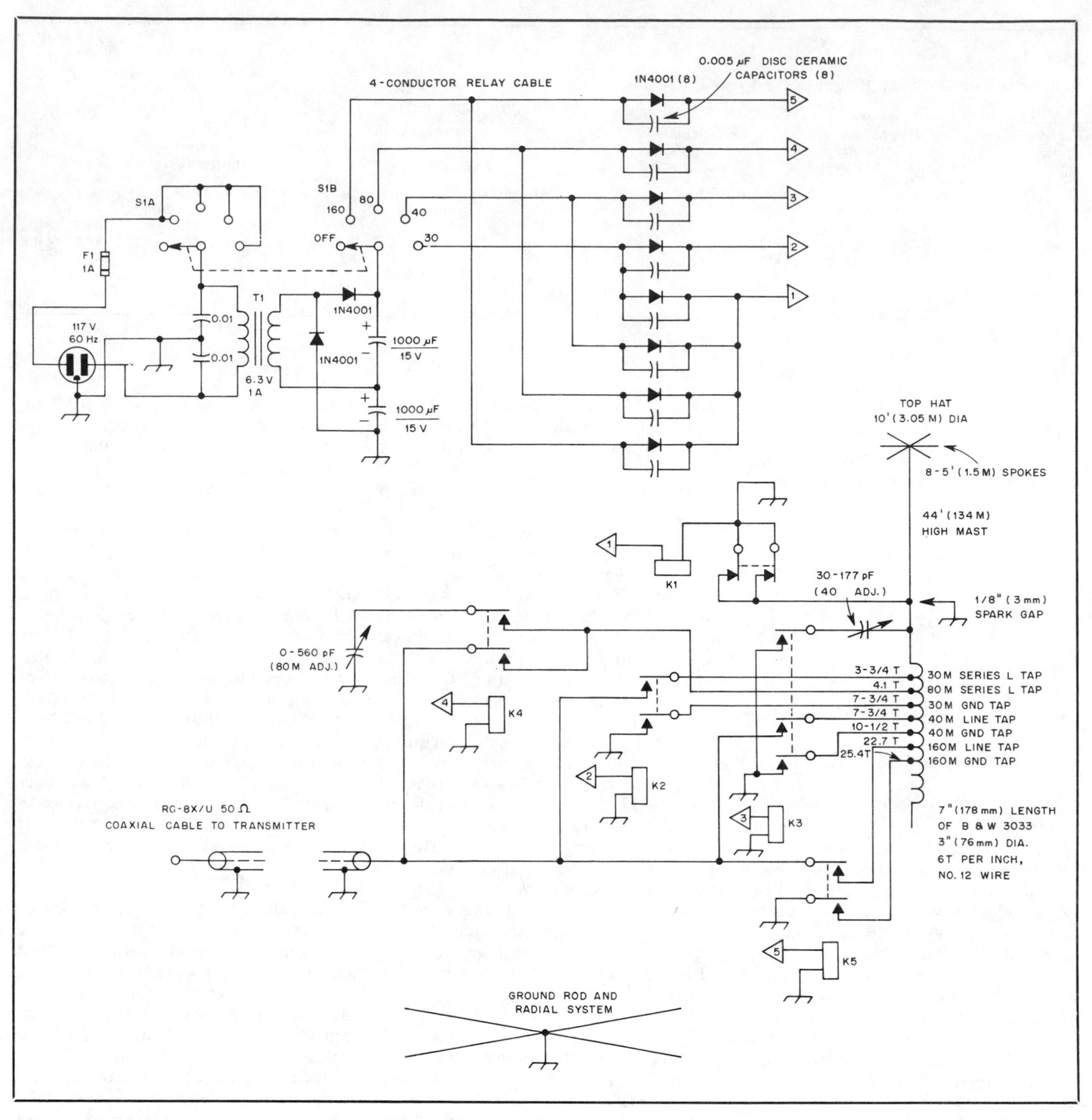

Fig. 1 — Schematic diagram for the K3EQ 160-, 80-, 40- and 30-meter vertical antenna. The circuit for remote band switching is included. There are 52 radials and a ground rod in the system. Low-angle radiation makes this an effective DX antenna.

and a stainless-steel hose clamp. All four corners of the aluminum plate are connected to the ground feedthrough in the bottom of the tuning unit with heavy copper braid. This insulator, as well as the one for the lead going to the base of the mast, is sealed against moisture by applying silicone compound. The relay control cable and the 50-ohm coaxial line enter the bottom of the tuning box through small holes that ensure a snug fit. The completed antenna, as shown in the photograph, has the capacitance hat resting atop the mast. The mast is stabilized by careful positioning of the guy wires. A wooden fence is placed around the base of the mast to help protect people and animals from possible rf burns.

Radial System

Installation of the mast took place during the driest Pennsylvania summer in 15 years. As fall approached, the soil was still too hard to bury the radials, so they were laid on the surface. Each wire was stretched tightly and fastened with several 6-in. lengths of heavy bus wire, which had been formed into hooks. When rain eventually fell, the radials were buried 2 to 3 in. in the ground.

All radials are 100 ft long except those toward the sides of the lot (which is only 150 ft wide). One side has 70-ft radials, while the other has 80-ft radials. Some 4800 ft of wire makes up the 52 radials. I used insulated hookup wire, but aluminum clothesline[10] or galvanized electric fence wire is satisfactory.

According to Stanley,[11] the efficiency

Fig. 2 — A view of the vertical antenna tuning network. Components are mounted on a framework of 1/4-in. thick Plexiglas, which slides into the fiberglass box.

Table 3

Shopping List

1 — telescoping TV mast, 40 ft long, Montgomery Ward no. 63A19735R, $39.95.
1 — galvanized fence post, 6 ft long, 2-in. dia., $6.
4 — lengths of thin-wall conduit, 10 ft long, 1/2-in. dia. Each length is to be cut into 5-ft sections. Montgomery Ward no. Z83A1004R, size no. 2, $1.89 ea.
1 — length of 3/4-in. galvanized water pipe for ground rod, 10 ft long, Montgomery Ward no. 81A40103R, $12.
2 — rolls of no. 12-1/2 gauge galvanized steel wire for guys, Sears no. 32H10125, $6.29 ea.
3 — earth anchors, screw type, 48-in. long, Sears no. 32H21946C, $7. ea.
42 — strain insulators for guy wires, Radio Shack no. 270-1518. Price with 10% quantity discount, $13.04.
120 ft (36.5m) RG8X-50 coaxial cable available from Texas Towers, Plano, Texas, $18.
120 ft four-conductor control cable for relay circuit, gray vinyl jacket. Sold by Fair Radio Sales, Lima, Ohio, $14.40. A substitute would be TV rotator cable, Sears no. 57H6732, 10¢ per foot.
5000 ft no. 18 vinyl-covered hook-up wire for radials, sold by Fair Radio Sales, $75. A less expensive (but less durable) substitute is no. 17 gauge galvanized steel wire. This is available from Sears, no. 32H22056C, at $16 per roll. Each roll has 2640 ft of wire.
1 — B & W coil no. 3033, 10 in. long, 3-in. dia, no. 12 wire, 6 tpi, available from Barker and Williamson, 10 Canal St., Bristol, PA 19007, $7.97.
1— fiberglass case, 14-1/2 × 14 × 4-1/4 in., available from Fair Radio Sales, $5.
5 — relays, dpdt plus spst, N.O., 12 V dc, Leach no. 1077, available from Fair Radio Sales, $2 each.
1 — variable capacitor, 30-177 pF with both sections in parallel, 0.094-in. air gap. Fair Radio Sales, no. C-221/T-195, $3.95.
1 — variable capacitor, 0-563 pF, 0.03-in. air gap, Fair Radio Sales, no. 76348-C, $2.95.
2 — cone-style feedthrough insulators, Fair Radio Sales, no. 3G5841N-84, 25¢ each.
1 — standoff insulator, 7-in. × 1-1/4 in. dia, Fair Radio Sales, no 5970-405-8992, $4.

Miscellaneous: parts for control box purchased from Radio Shack, $20.
Stainless-steel hardware from Elwick Supply Co., Somerdale, New Jersey, $12.
Aluminum angle stock and 1/8-in. aluminum plates from local metal suppliers, hose clamps, ready-mix concrete, copper shielding and braid, $10.

Note: The total cost was approximately $300 at the time the antenna was built. It is reasonable to expect the present costs to be about 10% higher. By "scrounging" parts from your junk box, and from friends and flea markets, the cost can be reduced.

of a 160-meter antenna might be improved by using more or longer radials. For the other bands, however, not much improvement is likely to be achieved by increasing the lengths or adding radials. For 160 meters, the radials are only 0.184 wavelength, but for 80 meters they are a respectable 0.352 wavelength long. Ground losses are probably on the order of 2 dB on 80 meters and about double that on 160 meters. Table 4 is a chart of the wavelengths of the 100-ft radials versus frequency.

Tuning

Tune-up is done on 160 meters first, then progressively on the higher bands. I used the K4KI[12] tune-up bridge and a dummy load at the base of the antenna to make the adjustments. My transmitter was in the second-floor shack. I should have carried it to the base of the antenna to make the matching process easier. Finding the correct coil taps for 160 meters while using the bridge seemed almost impossible. By tightly coupling a grid-dip oscillator to a two-turn link in the ground lead, the correct ground tap point was located. The line tap was then positioned properly with the aid of the tune-up bridge. Adjustments for the other bands followed without difficulty. The required inductances were close to the calculated values. Fig. 4 shows an SWR plot for the antenna. Refer also to Table 5.

This data was obtained in the shack at the end of the 120-ft length of RG-8X coaxial feed line. The SWR might be brought closer to 1:1 on 30 meters by further adjustments for that band. After the tap points on the coil were found, I soldered miniature alligator clips to the coil. A purist might prefer to remove the

Fig. 3 — Base of the vertical antenna with the tuning-component box mounted on the ground rod. The radials terminate on a square aluminum plate.

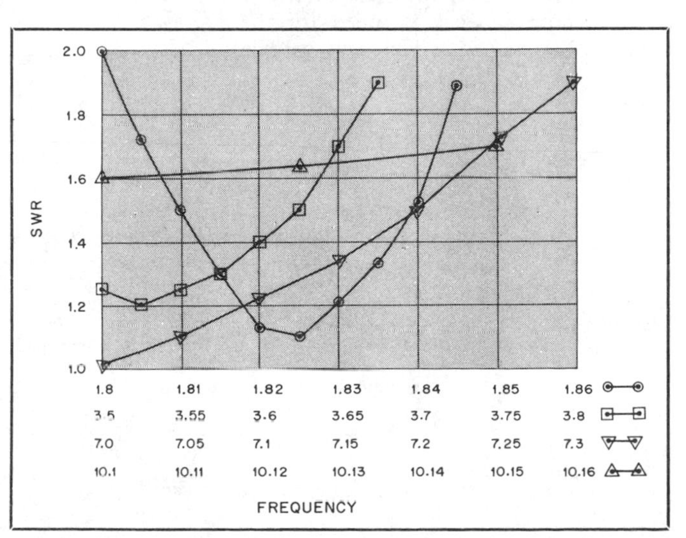

Fig. 4 — SWR curves for the K3EQ vertical antenna. See Table 5 for related information.

clips and solder the braid directly on the coil. I left the clips there to facilitate future adjustments.

The antenna is resonant outside the low ends of the 30- and 40-meter bands. This apparently results from the extra foot or so of wire from the base of the mast to the tuning box and ground. Additionally, I did not cut the "top hat" to the calculated 9.6-ft diameter, but left it at 10 ft. Shortening the mast 1 foot should bring the resonant points within the 30- and 40-meter bands. If additional correction is needed, then remove 2.5 in. from each of the "top-hat" spokes. This change may require repositioning of the taps from the points indicated on the schematic diagram (Fig. 1).

Afterthoughts

Phone operators may think this article has nothing to offer them. Therefore, I went through an exercise to determine the optimum configuration to cover the new 30-meter band and the 160-meter band, and also to allow adjustment for the lowest SWR at the center of the 40- and 75-meter phone bands. To accomplish this, the mast must be lengthened to 47.5 ft, and the top-hat diameter reduced to 5.8 ft. Table 6 charts the calculations that lead to this conclusion.

Of course the tuning network would allow this configuration to be tuned to the 40- and 80-meter cw bands by those operators who might like to tune the antenna to any part of these bands. For 80-meter cw, more series inductance would be needed for the 44-ft version. For 40-meter cw, some series inductance would have to be inserted between the mast base and the parallel-tuned tank. This requires only moving all three 40-meter coil taps down the coil a few turns. Proper adjustment for operation anywhere in the 40- or 80-meter bands can be made with this configuration.

The full 40-meter band could be covered with an SWR of 1.4:1 or less if this matching network is tuned for the lowest SWR at 7.15 MHz. This can be verified by extrapolating the SWR curves of Fig. 4. Likewise, it appears that if the configuration were tuned for the lowest SWR at 3.8875 MHz, all of the 75-meter phone band could be covered with an SWR of 1.7:1 or less.

Conditions were not favorable for evaluating its DX qualifications when I conducted tests with this antenna. Results obtained were nevertheless gratifying. Europe and South America have been worked with very good reports on 80 and 40 meters. On 160 meters, with 100-watts input to a TX4C, I received an RST 589 report from KP2A followed by a 549 from VP9KA. To the west, my circle of contacts has been from Minnesota (559) through Wisconsin (579), Iowa (559), Kansas (539) and Arkansas (559). A 339 report came from New Mexico, and a station in Florida gave me a 579. All of these contacts were made in the early evening.

I have shown the math calculation results in tables. Readers may get the calculations from ARRL. Send an SASE with $1 to the Technical Department Secretary and request the 9/81 *QST* "Antenna Design and Math Calculations."

If you wish to enhance your DX capabilities on the lower bands without erecting a "monster antenna," to be prepared for the new 30-meter band when it becomes available or to try the recently expanded "top band" for the first time, then this may be just the antenna for you. Build it, and you'll be ready for some good DXing!

Table 4

Length of Ground Radials in Wavelengths Versus Frequency

F (MHz)	*100-ft (30-m) Radials (length in λ)*
1.8125	0.184
3.525	0.357
7.025	0.712
10.125	1.029

Notes

[1]meters = feet × 0.3048.

[2]D. Hollander, "A Big Signal from a Small Lot," in this chapter.

[3]Editors of *73 Magazine, The Giant Book of Amateur Radio Antennas* (Summit, PA: Tab Books).

[4]J. Sevick, "The W2FMI Ground-Mounted Short Vertical," *QST*, March 1973, pp. 13-18, et al.

[5]J. Sevick, "Short Ground-Radial Systems for Short Verticals," in Chapter 6.

[6]W. Schulz, "Designing a Vertical Antenna," in Chapter 1.

[7]millimeters = inches × 25.4.

[8]See note 2.

[9]See note 4.

[10][Editor's Note: In regions where the soil has a high acid or alkaline content, rapid disintegration of aluminum wire will occur, sometimes within a few months. Neoprene-jacketed no. 8 aluminum wire (sold by Sears as overhead power wiring for outdoor applications) is relatively inexpensive and is highly resistive to corrosion.]

[11]J. Stanley, "Optimum Ground Systems for Vertical Antennas," *QST*, Dec. 1976, pp. 13-15.

[12]W. Vissers, "Tune Up Swiftly, Silently and Safely," *QST*, Dec. 1979, pp. 42-43.

Table 5

Data for SWR Curves in Fig. 4

Frequency (MHz)	*SWR*
1.8	2.0
1.805	1.72
1.81	1.5
1.815	1.3
1.82	1.13
1.825	1.1
1.83	1.21
1.835	1.33
1.84	1.52
1.845	1.89
3.5	1.25
3.525	1.2
3.55	1.25
3.575	1.3
3.6	1.4
3.625	1.5
3.65	1.7
3.675	1.9
7.0	1.01
7.05	1.1
7.1	1.22
7.15	1.34
7.2	1.5
7.25	1.72
7.3	1.9
10.1	1.6
10.125	1.64
10.15	1.7

Table 6

Chart for Selecting Optimum Radiator for Phone Bands (7.225 and 3.8875 MHz)

Radiator Height (ft)	*Top Hat Cap (pF)*	*Top Hat Dia (ft)*	*Calculated Heights for 10.125 MHz (Deg.) Sum — 225 Deg.*		*Calculated Heights for 7.225 MHz (Deg.)*			*Calculated Heights for 3.8875 MHz (Deg.)*			*Calculated Heights for 1.8125 MHz (Deg.)*		
			Height	Top Loading	Height	Top Loading	Sum	Height	Top Loading	Sum	Height	Top Loading	Sum
47	49	6	174.3	50.7	124.39	41.08	165.47	66.93	25.13	92.06	31.2	12.27	43.47
47.5	46	5.8	176.17	48.83	125.7	39.2	164.9	67.64	23.69	91.33	31.54	11.56	43.1
48	43	5.3	178.02	46.98	127	37.36	164.36	68.35	22.33	90.68	31.87	10.84	42.7
49	37.7	4.75	181.73	43.27	129.68	33.88	163.56	69.78	19.86	89.64	32.53	9.56	42.09

Note: Subtract length of lead into tuning unit plus ground lead from calculated radiator height.

from April 1979 *QST*

A Big Signal from a Small Lot

Good things often come in small packages. Consider N7RK's 60-foot vertical, a proven top-notch DX contest antenna.

By David S. Hollander,* N7RK, ex-WB6NRK/7

Regardless of what others throughout the rest of the country may think, not every radio amateur in Arizona has 10 acres of land or more on which to farm exotic antennas for DXing. On the contrary, indeed, many of us in the Grand Canyon state reside in apartments or homes with little or no space for outdoor antennas. Where there *is* space, most likely it is insufficient for installing an aerial designed for use on 75 or 80 meters, much less 160.

* 2313 E Ocotillo Rd, Phoenix, AZ, 85016

Among these city dwellers living in the shadow of such limitations are many amateurs, like myself, who prefer operating on the low-frequency bands. Some, perhaps, have resigned themselves to the facts of their individual lives and settled for operation on 10, 15 or 20 meters or the uhf bands. Although I was faced with similar restrictions, dismay was not about to rule me out of my favorite bands. In the end I erected an antenna that rewarded me with a gratifying amount of DX and helped me earn a position in the top brackets of the 1976 DX contest on 160!

The N7RK antenna loading coil and relay box. Transmitting-type air-wound coils such as used in this photograph may be found in surplus military gear, dismantled broadcast transmitters or may be obtained from G. R. Whitehouse & Co., 11 Newbury Dr., Amherst, NH 03031.

The long winter nights of 1975, during which I was living in an apartment in Tempe, AZ, gave me the opportunity to observe that DX-minded amateurs on 75, 80 and even 160 were communicating across the oceans with modest amounts of power and, seemingly, with little difficulty. As I scanned these frequencies night after night, my desire to be a part of this DX action became more fervent. I realize that some *QST* readers may reason that if DX happens to be my goal, I should simply switch to 10, 15 or 20 meters. My ready-made rebuttal, of course, is that DXing on these bands lacks the challenge to be found on 75, 80 and especially 160 meters.

From Horizontal to Vertical

Conditions at the apartment where I lived that winter were such that I could install a 150-foot horizontal antenna. This end-fed wire, however, left much to be desired as far as DX is concerned. After all it was only 20 feet above ground. As a result the angle of radiation was very high. The best compliment I can give this antenna is that it did more to keep my interests aglow than anything else.

An observation I made that winter coincided with antenna theory dating back to the early days of radio. The best DX on 75, 80 and 160 meters came from stations having vertical antennas. This seemed particularly true on 160. Of course, an exception might occur if an amateur station had a horizontal antenna suspended 120 feet above ground. In a practical sense, however, such a situation is quite unlikely for the amateur whose residence is on a small city lot.

In time, I moved from the apartment to a house. While this transition gave me more personal freedom to work with antennas, the lot size, just 60 × 25 feet (18.3 × 7.6 m), offered no advantage for stringing up a suitable wire antenna for my favorite bands. No longer could I have a 150-foot antenna. In fact the smallness of the yard precluded the erection of even a half-wave antenna for 40 meters. Nevertheless, I was not dismayed by the prospects of being unable to string out a skywire.

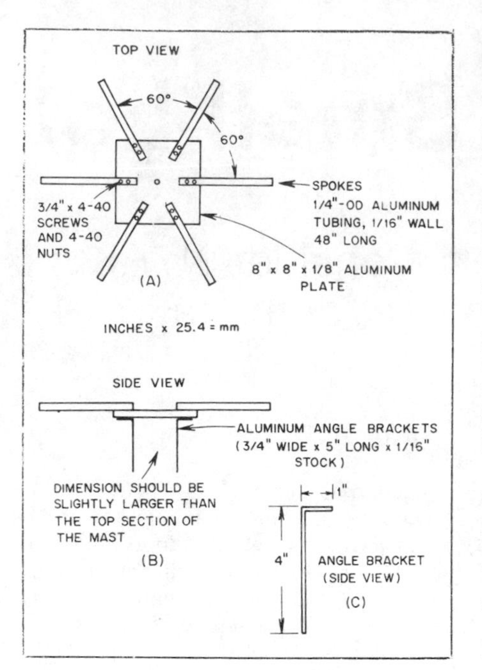

Fig. 1 — Construction of the N7RK vertical antenna capacitance hat. Each spoke is fastened to the aluminum plate by means of 3/4 × 4-40 machine screws and 4-40 nuts. The capacitance hat is fastened to the mast by means of plumbing or automotive hose clamps. To obtain good continuity between the hat and the mast, a length of wire mesh braid should be brazed to the mast and secured to the aluminum plate with the use of a solder lug and machine screw.

Long before leaving the apartment, I had firmly decided that my next antenna would be vertical. For that reason, the dimensions of the lot were of minor concern. Anxious to be ready for the next DX season, I experienced no difficulty in being motivated to erect a 60-foot radiator. Preparations began soon after getting settled at the new location.

That I decided upon a height of 60 feet was largely a matter of the material on hand. For some time I had saved some tubular TV masts, thinking that the day would come when they would be useful for my Amateur Radio activities.

Obviously there was little choice in where to locate the antenna. It could be erected in the middle of the yard or close to the house. My decision was to forgo any technical advantage of having the mast in the middle of the yard in order to leave that area free of any encumbrances. Instead it now stands against the carport. Placing the mast at this location enabled me to secure it to one of the 4 × 4 supports for the carport roof. Clamps placed at the 3- and 8-foot levels hold the mast firmly against the wooden upright.

About Construction

The material on hand consisted mainly of a 50-foot (15.2-m) telescoping TV mast. This turned out to provide a length of 44 feet when assembled. Consequently an additional 16-foot section was needed.

A capacitance hat, illustrated in Fig. 1, furnishes additional electrical length. The hat has six 4-foot tubular aluminum spokes equally spaced atop an aluminum plate. Each spoke is fastened to the plate with machine screws.

The mast is guyed at the 35-foot level and a point 2 feet below the top. I used ordinary TV guy wire, such as one may obtain at most electronics stores. To prevent any unwanted resonances occurring along the guys, each wire is broken at random points with insulators installed between the segments. Turnbuckles provide means for tightening the wires. Anchor points are each 25 feet from the base (two on the house and one on a fence post).

Purists might cast a jaundiced eye at my method of supporting the bottom of the mast above ground. Lacking a suitable insulator, I simply placed it atop a cinder block. No appreciable loss seems to have resulted, however.

The Ground System

A ground rod alone is usually not considered to be an effective means of providing a ground for a vertical antenna. Explanations to this effect are found in *The ARRL Antenna Book* and other texts. The ideal ground system would have been 120 radials, at least 1/2-wavelength long, spaced equally around the base of the antenna every 3 degrees.[1] Inasmuch as the distance from the base of the antenna to the end of the yard was only 60 feet, such an installation was slightly out of the question. Instead, I used far less than that amount. Indeed, amateurs often obtain good results with far fewer radials. My situation is a case in point.

Because of the space limitations at my new home, my ground system had to be far short of the ideal. I compromised by laying out 20 radials of no. 12 copper wire, 60 feet in length and buried 3 inches in the ground.

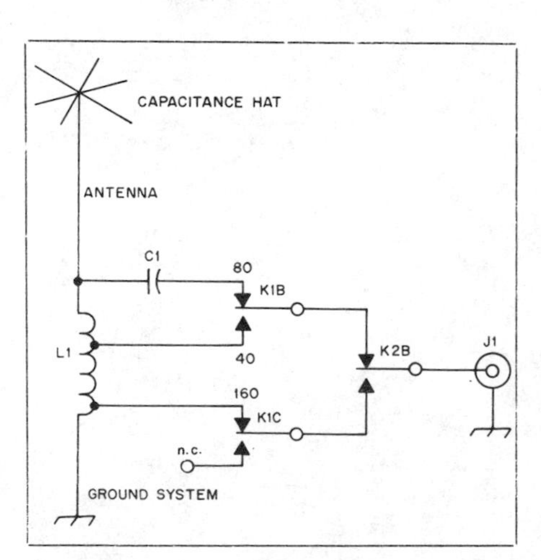

Fig. 2 — Antenna matching system for the N7RK vertical. The ground system consists of 20 radials of no. 12 copper wire, each 60 ft long (see text). L1 is a B & W no. 3035 coil tapped at 7 turns from the top and 4 turns from the bottom. Four 1200-pF fixed capacitors in series are used for C1. K1 and K2 are Potter and Brumfield KRP11DG dpdt 12-V dc relays with 10-A contacts. Only one set of contacts is used for K2. J1 is an SO-239 uhf connector.

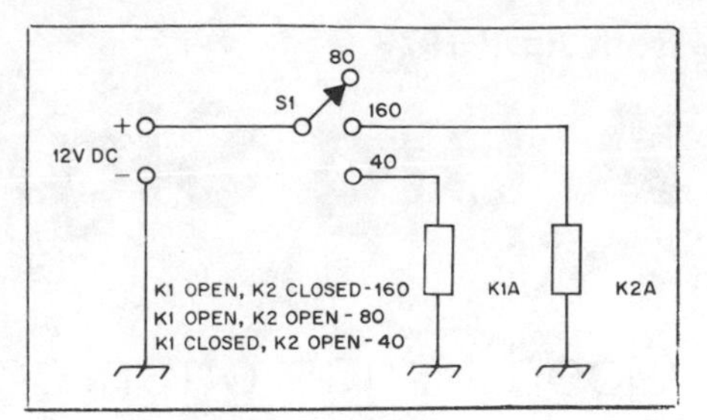

Fig. 3 — Band changing of the N7RK vertical antenna system is simplified by this relay control system. A two-conductor control cable to the remote outdoor tuning system is required. The ground return is by way of the coaxial feed-line braid. S1 is a single-pole, three-position rotary switch located on a control box in the station.

After reading further on the subject of ground systems,[2,3] I learned that if only a small number of radials are used, there is little point in extending them out 1/4 or 1/2 wavelength. I reasoned that 1/8 wavelength would have been adequate for my installation. According to the literature I read, a large number of radials, even though short, are preferable to a few long radials. Most of the ground losses seem to occur near the base of a vertical antenna. Therefore, within reason, the more metallic surface area a radial system has near the base, the lower will be the ground losses.

Antenna Matching

In order to obtain a good antenna match for each of the bands, I relied mainly on a dip oscillator, an SWR indicator and a bit of cut and try. Arbitrarily, I decided to adjust the antenna first for operation on 75 meters. With the oscillator coupled to the base of the mast, I noted a dip that occurred at 3 MHz. This also indicated that the antenna was approximately 1/4 wavelength long at that frequency. However, at 3.8 MHz it would appear to be longer than 1/4 wavelength, with inductive reactance being evident at the feed point. To cancel this reactance, I inserted a capacitance of 300 pF in series with the antenna, as shown in Fig. 2. I arrived at that value purely by experimentation. Four 1200-pF, 500-V dipped-mica capacitors wired in series provide the 300-pF capacitance.

A 60-foot vertical antenna will display capacitive reactance at the feed point when operated on 160 meters, because it is electrically shorter than 1/4 wavelength. To cancel this capacitive reactance, inductance must be introduced into the antenna circuit. Fig. 2 shows this inductance, L1, in series between the antenna and the ground. Taps on the coil enable it to be used for both 160 and 40 meters. Although the tap points I use for L1 are indicated in the caption for Fig. 1, they

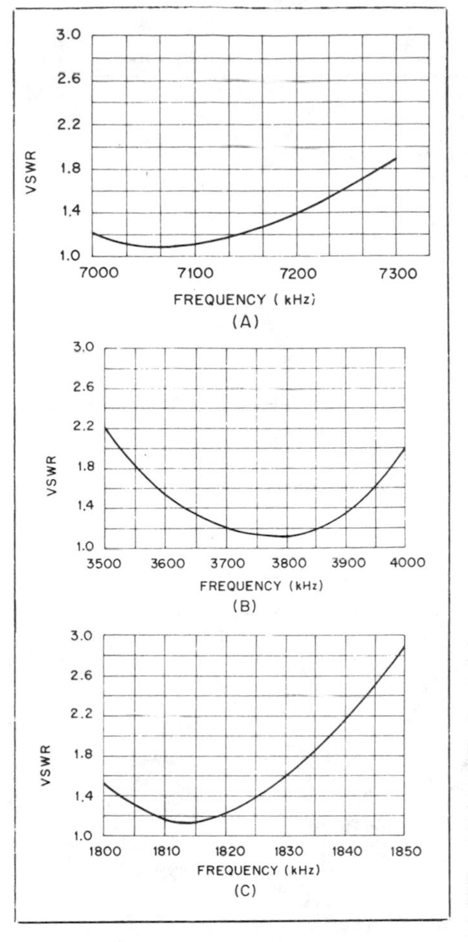

Fig. 4 — The three graphs compare the VSWR versus frequency for the N7RK vertical antenna for the 40- through 160-meter amateur bands.

This close-up photograph shows how the antenna changeover relays are housed at the base of the N7RK vertical antenna. The relays, remotely controlled from the shack, provide quick band changing.

may not be the same at other installations. Proper tap adjustment and minimum SWR are obtained rather easily by experimentation. Fig. 4 shows the results of SWR measurements at N7RK.

Control System

The simple control system illustrated in Figs. 2 and 3 eliminates trips to the antenna each time a switch from one band to another is made. L1 is a Barker and Williamson no. 3035 inductor. If one is not available, a suitable substitute would be an air-wound coil 2 inches in diameter (51 mm) having 40 turns. Such inductors are often found at flea markets or are offered for sale by surplus dealers. These same sources may also have a relay that is equivalent to the Potter and Brumfield 12-V unit no. KRP11DG that I use. A good idea is to check the coil voltage rating. For example, Potter and Brumfield makes this model with coils for 6, 12, 24, 48 and 110 V dc.

The relays should be protected from the weather by a suitable housing. I chose not to enclose the inductor perhaps because of the generally favorable weather conditions in my area. For regions subject to much moisture, enclosing both the coil and the relays seems desirable. Plastic freezer boxes are relatively inexpensive yet quite satisfactory for this purpose.

Only a two-wire cable from the shack to the changeover relays is required for the control circuit. The coaxial-cable braid may be used for the ground return. My relays are operated by a homemade 12-volt unregulated supply. Relays of other voltage ratings may be used, of course, but in such cases the supply voltage and current must be compatible with the particular relay to be used.

Antenna Performance

As evidence of success with this vertical antenna, let me offer these results. In the 1976 ARRL 160-meter contest I logged 345 QSOs and had 73 multipliers and 48 states. That earned me the second-highest score in the West and made me the winner for the state of Arizona. Furthermore, I established ocean-hopping contacts with Japan and New Zealand. I've worked ZL2BT several times on phone and cw. I should also point out that all of these contacts were made with just 200 watts input.

The success I've experienced with this antenna didn't end there. Over a six-month period I worked 97 countries on the 75-meter band. Many European stations were worked, even though that part of the world is generally difficult to contact from this section of the United States. An advantage I had on 75 meters, I will admit, is that I used a full kilowatt input.

Because the antenna also performs well on 40 meters, I've had rewarding experiences on that band, too. Not only do my 40-meter contacts include European stations but also stations deep in Asia. I have log entries for call signs such as VU2, 4S7, UJ8 and VQ9, all worked on the long path according to my friends with rotary antennas. They told me which direction the signals were from since I couldn't rotate my vertical!

If you have little room to install a typical wire antenna for the lower frequency amateur bands, consider the results I've obtained with this antenna. The world could be at your fingertips with the help of a simple 60-foot stick.

Notes

[1] "Grounding Systems," *The ARRL Antenna Book,* 13th edition, 1974, p. 61.
[2] See note 1.
[3] Stanley, "Optimum Ground Systems for Vertical Antennas," *QST,* December 1976.

The Skeleton Discone

By D. Wilson Cooke, WA4RHT
PO Box 203
Tigerville, SC 29688

Do you need a coax-fed, broadband antenna that works on several amateur bands with little or no adjustment? This article describes experiments in which a discone antenna was stripped to a simple skeleton. These experiments suggest that a skeleton of a discone antenna can be constructed of wire, rod or tubing. It will work over several amateur bands without adjustment, and will maintain an SWR in a 50-Ω line of 2:1 or less. Sufficient data is given so that the element lengths can be calculated for any reasonable antenna frequency. Step-by-step construction details, however, are not provided.

The discone antenna was introduced in the mid '40s[1] and has been described infrequently in the amateur literature. Unlike dipoles, verticals, quads, Yagis and other popular antennas, the discone is not as well known or as widely used. The 15th (1988) edition of *The ARRL Antenna Book* contains a section on discone antennas for both HF and VHF use.[2] This material is recommended reading for anyone wishing further information on this unusual antenna.

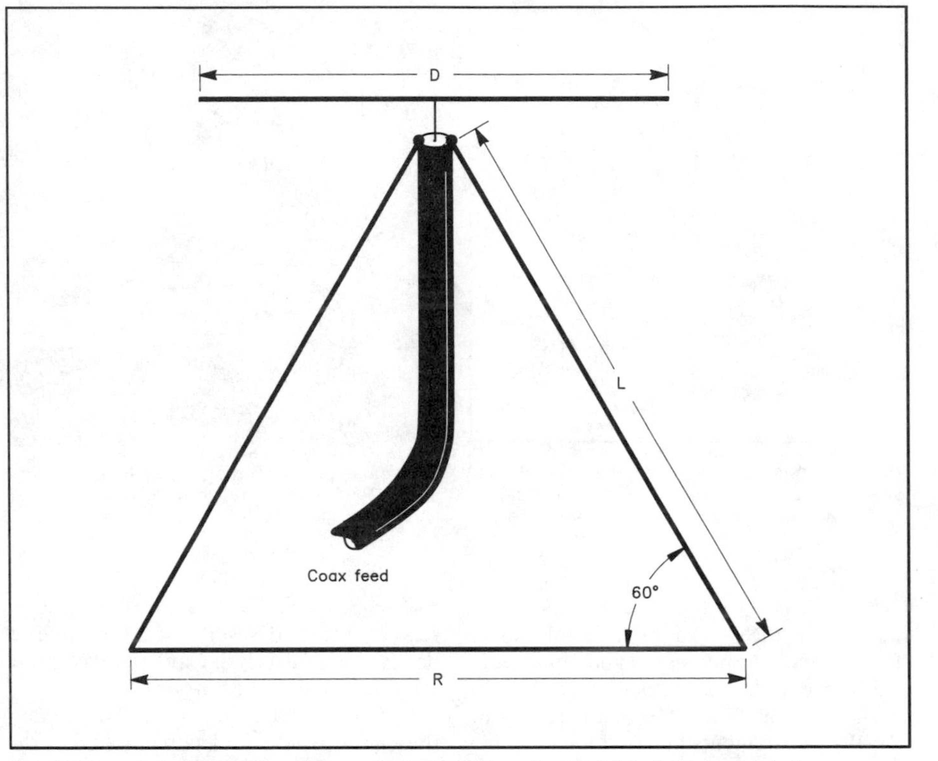

Fig 1—Cross section of the discone antenna. The slant height of the cone, L, is a quarter wavelength at the lowest operating frequency. The diameter of the disk, D, is 67-70% of L. The shield of the coax feed is attached to the top of the cone and the center conductor is attached to the center of the disk.

Fig 1 shows a cross section of the discone antenna. The bottom of the antenna consists of a metal cone whose slant height, L, is equal to the diameter of the base of the cone, R. Above the cone is a metal disk with a diameter (D) equivalent to about 67% to 70% of the slant height of the cone. The coax feed-line shield is attached to the top of the cone while the center conductor passes through a small hole in the top of the cone and is attached to the disk.

The slant height, L, is a free-space quarter wave ($246/f_{feet}$ or $75/f_{meters}$, where f is the frequency in MHz) at the lowest frequency for which the antenna is designed. Below this frequency the SWR increases rapidly. If the hole at the top of the cone is small with respect to the slant height, the antenna will provide a good match to 50-Ω coax over a frequency range of up to ten times the low-frequency cutoff.

As far as the feed line is concerned, the discone appears as a properly terminated high-pass filter. Radiation is vertically polarized at low angles over the lower frequencies. As the upper frequency limit of the antenna is approached, the radiation angle turns upward. Common vertical antennas like the groundplane and grounded vertical have the high current portion of the antenna at the bottom. The discone has the high current portion of the antenna at the top. The discone requires no ground or ground screen for its operation.

The size of a VHF discone is small enough to make construction and mounting of the antenna fairly easy. For example, an antenna with a low-frequency cutoff of 144 MHz might have a cone slant height of 20 inches and a disk diameter of 14 inches. In fact, a carefully constructed discone antenna might work through the 1.2-GHz band! It could be made of sheet aluminum or other metals.

A discone for HF is more difficult to construct. At 14 MHz the cone would require a slant height of at least 18 feet and a disk diameter of 12 feet. The construction and wind loading of such a discone would make the antenna impractical for most amateur applications.

Several skeletons of the discone have been made in which the disk and cone are simulated by wires or rods. Radio Shack and Procomm market similar VHF skeleton discones. The disk is replaced by eight horizontal spokes about one foot in length, giving a "disk" diameter of two feet. The cone is simulated by eight rods tilted down at a 60° angle similar to drooping radials. These are about three feet in length. The low-frequency cutoff of skeleton discone antennas is approximately 100 MHz. Above this frequency the SWR is usually below 1.5:1 well into the UHF region. Below 100 MHz, however, the SWR rises rapidly.

The ARRL Antenna Book describes a high-frequency skeleton discone with a low-frequency cutoff of 7 MHz. The "disk" consists of eight horizontal aluminum spokes and the "cone" is simulated by 24 Copperweld wires. (The wires also act as guys for the tower that supports the structure.) The tips of both the disk and cone spokes are connected by a wire loop. The *Radio Handbook* describes a similar antenna except that the cone is simulated by 48 wires.[3]

I became curious about the performance differences between various skeleton discones and full sheet-metal discones. I also wanted to know the minimum number of spokes in the skeleton that would yield acceptable performance.

The Skeleton Experiments

It is relatively simple to model an antenna at VHF and then scale results to lower frequencies. I chose the middle VHF region as the principal frequency for investigation because of equipment availability.

Several skeleton discones were constructed using different numbers of spokes to simulate the disk and the cone. I selected a "cone" slant height (spoke length) of 38 inches and a "disk" diameter of 25.5 inches. These dimensions were selected because they were small enough to manage easily in testing, and yet large enough that cutting and measuring errors would account for a small percentage of the overall size of the antenna. I estimated that the low cutoff frequency point would be between the 6-meter and 2-meter amateur bands.

The foundation for the antenna was an aluminum funnel with a BNC connector mounted at the small end to serve as the junction between the disk and cone portions of the antenna. A machine screw was soldered to the center conductor of the BNC connector and it projected out of the top of the funnel. The spokes simulating the cone

Fig 2—One of the skeleton discones discussed in the text. Eight spokes simulate the cone and four simulate the disk.

could then be quickly attached to the funnel and various disk-simulating spokes could also be attached to the machine screw. This arrangement would not be suitable for a permanent antenna, but it proved adequate for a temporary series of experiments. The experimental antennas were all connected to a 20-foot length of RG-58 coax leading to the measurement equipment.

Before discussing the actual measurements, a word should be said about measurement equipment. Like many amateurs, I do not have access to laboratory-grade equipment. Some of my measurement equipment is home constructed and calibrated, and some of it is low cost commercial equipment. I used a dip meter as a radio-frequency oscillator. It was coupled to a counter and SWR meter through a loop on a short length of coax. The output at the SWR meter was quite small. Consequently, the reflected power level was often in the range of a few millivolts. This means that the diodes in the SWR meter were probably operating in a nonlinear fashion. This would give rise to errors in SWR measurement.

Since all measurements were taken at approximately the same power levels while using the same equipment, the *relative* comparisons should be valid even though the absolute SWR measurements may somewhat inaccurate. All measurements were made in the yard with the antenna mounted on a wooden pole and the disk about seven feet above the ground. Several hundred measurements were made over a period of two weeks on a variety of different antenna configurations. I measured the SWR at 10-MHz intervals from 70 MHz to 320 MHz (the upper frequency limit of my dip meter). I tried configurations of eight, six, four and three spokes in various combinations. The results of these measurements are summarized in Table 1. Many other measurements were made that are not reported, but they were consistent with those in Table 1 and would contribute little additional information.

While a sheet-metal discone has a low-frequency cutoff when the slant height of the cone is approximately a quarter wave, the skeleton discone's low-frequency cutoff is significantly higher. The slant height of 38 inches used in these experiments should have given a cutoff at 77 MHz in a traditional discone, but the SWR was well above 3:1 at this frequency for most skeleton designs I attempted. It did not drop below 2:1 until the frequency reached 100 to 110 MHz (depending on the antenna spoke configurations). This represents a slant height close to *one-third* wavelength for the low-frequency cutoff.

The first configuration I tried was a copy of the commercial Radio Shack and Procomm discones. Eight spokes were used in the cone and the disk. The tips of neither the disk nor the cone were connected. The SWR dropped below 2:1 just above 100 MHz and remained low up to 320 MHz (column A of Table 1). At some points the SWR was below 1.1:1 for several MHz and was below 1.5:1 over most of the range. It did rise slightly at certain frequencies with a maximum SWR of 2.1:1 at 310 MHz. This leads me to conclude that this design would provide a perfectly acceptable antenna for scanner use, as well as for amateur transmissions on the 144, 222, and probably the 420-MHz bands. (No measurements were made on the latter band.)

I substituted a four-spoke disk at the top of the antenna while leaving the eight-spoke cone unmodified. The SWR measurements are reported in column B of Table 1. The SWR profile is slightly different, being higher on some frequencies and lower on others. Even so, this configuration would be as acceptable as the eight-spoke disk.

Column C reports the SWR profile of an antenna with a four-spoke disk and a six-spoke cone. The SWR is generally higher, though still below 2:1 throughout most of the range above 110 MHz. Three peaks are reported in the table where the SWR rose above 2:1. None of these are within amateur bands.

Table 1
SWR Profiles of Several Skeleton Discones Discussed in the Text

Frequency	*Antenna SWR*				
(MHz)	*A*	*B*	*C*	*D*	*E*
70	4.9	4.8	5.8	4.8	5.7
80	2.7	3.2	4.5	4.1	4.2
90	2.0	3.4	4.4	4.4	3.7
100	1.7	2.7	6.6	4.1	4.2
110	1.7	1.7	1.9	1.7	1.9
120	1.0	1.3	2.6	1.2	1.5
130	1.1	1.3	1.3	1.4	1.5
140	1.0	1.1	1.1	1.8	1.2
145	1.2	1.3	1.5	2.1	1.5
150	1.4	1.3	1.5	2.0	1.5
160	1.2	1.3	1.4	2.2	1.5
170	1.1	1.2	1.3	1.7	1.3
180	1.1	1.2	1.8	3.2	2.0
190	1.2	1.4	1.4	2.1	1.3
200	1.2	1.2	1.5	2.8	1.6
210	1.6	1.6	2.0	3.0	2.0
220	1.1	1.0	1.1	1.6	2.1
225	1.5	1.6	1.5	3.7	1.6
230	1.1	1.1	1.2	2.7	1.1
240	1.4	1.5	1.2	2.5	1.3
250	1.2	1.9	1.1	1.2	1.8
260	1.5	1.4	1.2	1.2	1.4
270	1.2	1.4	1.2	1.2	1.4
280	2.0	2.2	2.5	2.1	2.6
290	1.6	1.8	1.8	2.1	1.8
300	1.3	1.5	1.4	1.4	1.7
310	2.1	1.9	2.4	2.2	2.4
320	1.4	1.6	1.6	1.7	1.3

A—Skeleton with eight spokes in the cone and the disk.

B—Skeleton with eight spokes in the cone and four in the disk.

C—Skeleton with six spokes in the cone and four in the disk.

D—Skeleton with four spokes in the cone and the disk.

E—Skeleton with six spokes in the cone and three in the disk.

Column D reports the SWR profile of an antenna with four spokes in the disk and in the cone. The SWR peaks are in the wrong places for the 144- and 222-MHz bands, though they are acceptable for many applications. When using my 10-watt, 144-MHz transmitter, the SWR was high enough to activate the SWR shutdown feature. (The transmitter output dropped about 15%.) If I were to build a permanent version of this antenna, I would change the lengths of the spokes slightly to move the high SWR portions of the antenna outside the amateur bands of interest. I believe a similar antenna could be constructed for general multiband use.

Column E reports the SWR profile on an antenna with three spokes in the disk and six spokes in the cone. This antenna would also be useful over the same frequency range.

Both *The ARRL Antenna Book* and the *Radio Handbook* describe HF skeleton discones in which the tips of the disk and cone spokes are connected by conducting wire. In order to determine what effect this would have on the antenna, I constructed a version of the discone with eight spokes in both the cone and the disk. All spoke tips were connected by circular wire rings.

This proved to be a mixed blessing. I did not run a complete SWR profile, but performed spot checks on several frequencies instead. The low-frequency cutoff dropped to about 77 MHz, but the SWR profile was generally higher. By utilizing the connecting rings, a physically smaller antenna could be made for the same low-frequency cutoff point. However, this approach may cause a slightly higher SWR at some frequencies and additional construction problems.

I tried a four-spoke cone/three-spoke disk version with circular rings connecting the spoke tips. It was totally unacceptable. The SWR throughout most of its range was between 2:1 and 3:1 with occasional dips and peaks.

Armed with the information outlined above, I constructed an HF version of the skeleton discone. I used four spokes, each six feet long, for the disk. I used four wires, each eighteen feet long, to simulate the cone. The top of the antenna was mounted on a push-up mast at about twenty feet. The wires were anchored to the ground with strings at an angle of about 60 degrees. The antenna was fed with 60 feet of RG8X coax. The SWR profile of this antenna is given in Table 2.

Table 2
SWR Profile of the HF Skeleton Discone with Four Spokes in the Disk and Cone

Frequency (MHz)	*Antenna SWR*
18.07	2.5
18.16	2.4
21.01	1.4
21.1	1.3
21.2	1.3
21.3	1.3
21.4	1.3
24.93	2.4
28.1	2.3
28.3	2.2
28.5	2.2
28.7	2.3
28.9	2.3
29.1	2.3
29.3	2.3
29.5	2.3
29.7	2.3
144	1.7
145	1.8
146	1.4
147	1.2
148	1.2

The SWR is certainly higher than desired in some bands. It does not, however, make the antenna totally unusable. For testing purposes I used a Kenwood TS-830S transceiver which features 6146 tubes as final amplifiers. It loaded easily on all HF amateur bands from 18 MHz and up without a tuner. The antenna also worked on 2 meters with both a mobile and a hand-held rig. I suspect that a solid-state HF rig would require a tuner to avoid SWR shutdown.

SWR alone is not a measure of good antenna performance. After all, my dummy load has an excellent SWR curve well into the VHF region, but it is a lousy antenna! Some verticals have a good SWR only because of significant ground loses. In order to determine how well the VHF version of the skeleton discone antenna performed on transmission and reception, it was compared with a more traditional discone antenna.

A traditional discone was constructed using hardware cloth covered with aluminum foil. The slant height was about 22.5 inches giving a low-frequency cutoff well below 144 MHz. All of my tests indicated that it performed as expected.

With my traditional discone as a reference, I made field-strength measurements on each of the five skeleton models outlined in Table 1. At a distance of twenty feet (roughly 3 λ) there was no measurable difference in radiated field strength between the skeleton discones and the traditional discone on 2 meters. Two distant repeaters with marginal signals at my QTH were used for reception testing. I found no measurable difference in received signal strength between the traditional discone and the skeleton discones. All of these measurements were made at the same location in my yard with the disks approximately seven feet above the ground.

The HF version of the skeleton discone could not be conveniently measured against a similar solid discone. I could only compare it with more conventional antennas. Obviously, my triband beam at 41 feet outperformed the discone. In its favored direction the beam is about two to four S units better. Off the sides and back of the beam, however, the discone is frequently better. Marginal signals are much easier to copy with the beam. This is to be expected since the beam is twice as high and has substantially greater gain.

I desired a fairer comparison for the HF skeleton discone so I constructed some temporary ¼-λ verticals with ground planes about six feet above the ground. Both field-strength and reception measurements indicated that the skeleton discone was as good as the verticals—and often better. Any signal that was readable on one antenna was also readable on the other. When monitoring signals from the United States, South America and Europe, I noticed that the skeleton discone had an advantage of one to two S units on 15 meters. Fewer signals were heard on 17 meters, but the results were similar. On 12 and 10 meters not many signals were available when I was checking the antennas. With the few signals that I did monitor, the discone performed as well as the verticals. This included one signal from the Azores on 10 meters.

I measured transmitted field strength from the verticals and the skeleton discone at a distance of 175 feet. The meter had a vertically polarized pickup antenna about 6 feet long. On 17, 15 and 12 meters the skeleton discone had a 2- or 3-dB advantage. On 10 meters the vertical-antenna signal strength was about 1 dB stronger.

Conclusions and Suggestions

I arrived at several conclusions as a result of these experiments:

1) Experiments can be conducted by amateurs with simple and inexpensive instruments and still achieve meaningful results.

2) The skeleton discone is a practical antenna that could serve as a general-purpose, wide-band antenna. It certainly will not perform as well as a single-band gain antenna adjusted for optimum performance, but it should be quite useful for many amateur applications. Element lengths are not critical if the builder makes them sufficiently long while maintaining the disk diameter at about 67-70% of the total cone spoke length.

3) As the number of spokes in the skeleton is decreased, the SWR curve has more variations. This indicates that one should use as many spokes as possible. An HF antenna with just a few spokes should work, however. The number of spokes in the cone appears to be more important than the number in the disk. A simple Transmatch could reduce the SWR so that even the most sensitive solid-state rigs would be usable. The additional line loss that is present with an SWR in the 2:1 range is not generally a problem at HF if a very simple skeleton is desired. For VHF I would recommend a version with eight or more spokes to obtain a flatter SWR curve.

4) A skeleton discone that does not utilize connecting rings at the tips of the spokes has a low-frequency cutoff point where the length of the cone spokes approaches one-third wavelength. When designing a skeleton discone, a length slightly longer than one-third wavelength ($330/f_{feet}$ or $100/f_{meters}$) at the lowest operating frequency should be chosen for the cone spokes. The length of the disk spokes (measured from center to tip) should be about 0.34 times the length of the cone spokes. This will give a disk diameter of just over two thirds the slant height of the cone.

5) A skeleton discone constructed with connecting rings requires more spokes for acceptable performance. A skeleton discone with eight or more connected spokes has a low-frequency cutoff point where the spoke length of the wire simulating the cone is about a quarter wavelength ($246/f_{feet}$ or $75/f_{meters}$). Therefore, the length of the elements for a given frequency are shorter, but *more* elements are required.

6) Because there are peaks and valleys in the SWR curve of a skeleton discone, it should be possible to design an antenna with low SWR points within the amateur bands. This would require some cut-and-try testing, but could produce an excellent antenna for general use.

The author would like to hear from others who have experimented with this sort of antenna. An SASE would be appreciated if a reply is desired.

Notes

[1]G. Kandoian, "Three New Antenna Types and Their Applications," *Proc IRE*, Volume 34, Feb 1946, pp 70w-75w.

[2]*The ARRL Antenna Book*, 15th edition (The American Radio Relay League, 1988), pp 9-7 through 9-12.

[3]W. I. Orr, *Radio Handbook*, 22nd edition (Howard W. Sams & Company, Inc, 1981), pp 27.24-27.26.

from February 1977 *QST* (Hints and Kinks)

HY-GAIN 18AVT/WB VERTICAL ANTENNA MODIFICATION

The Hy-Gain model 18AVT/WB is a vertical antenna, designed for use on the 80- through 10-meter amateur bands. Operation on 80 meters is possible because a mobile-type 75-meter loading coil and whip are mounted at the top of the antenna. Bandwidth on 80 is narrow, and the operating frequency is selected by adjusting the length of the whip. When the whip is cut to the proper length for 75-meter phone operation, it is not possible to obtain a proper match on the cw end of the band unless the whip is replaced with one of the proper length. Rather than purchase a new whip assembly, I fastened a 12-inch long piece of No. 12 copper wire to the whip with two Fahnestock clips. This allows me to slide the wire up and down on the whip until the proper position is found, using an swr bridge connected at the feed point of the antenna. — *Al Skornicka, K8WXQ*

[Editor's Note: A more permanent method is to use small cable clamps to secure the added wire to the whip.]

from March 1989 *QST*

Flagpole J For 10 Meters

You don't have to hide your antenna—put it out in the open for everyone to see! Just disguise it . . .

By Jim Hendershot, WA6VQP
3810 Almar Rd
Grants Pass, OR 97527

It had to happen. Even though I was a dyed-in-the-wool VHFer, HF fever was bound to hit me sooner or later. Before Novice enhancement, HF was a fascinating place where my higher-class ham buddies would spend long and wonderful hours, while I could only think of CW on the 40-meter Novice band. Yuk! Not being very good on CW, I couldn't get excited about spending hard-earned bucks to buy an HF rig just to use it for a few minutes on CW.

Two things changed all of that. The first was, of course, Novice enhancement. The second was the Santa Maria swapfest. Father's Day saw me heading down to the annual event loaded with junk—determined to make a killing. Not only did I do well, but after much judicious haggling and some good advice from one of my HF buddies, I also came home with a nice, new . . . well, almost new . . . IC-745. Oh, boy! Ten meters, here I come!

"Not so fast, young man!" I could hear my wife say it before I even asked. "You're not putting up some ugly antenna on my house! No way!" Having had this reaction from my XYL to the installation of even small VHF antennas, I knew putting up something big for use on HF would really be a problem. Something different was in order, but what?

After making some discrete inquiries, I discovered that my wife was wishing for a replacement flagpole for the house. A few months earlier, the old one had become unserviceable. That was it! A 10-meter antenna disguised as a flagpole! I immediately thought of using a quarter-wave antenna, but it was a bit too short for a good flagpole, and I couldn't figure out how I was going to hide all those radials.

I'd just been working on some antennas for the 2-meter band and 220, and had had great success with the J design. "Why not a 10-meter J?" I thought. Why not, indeed? So I set about building my wife's new flagpole—and my new antenna.

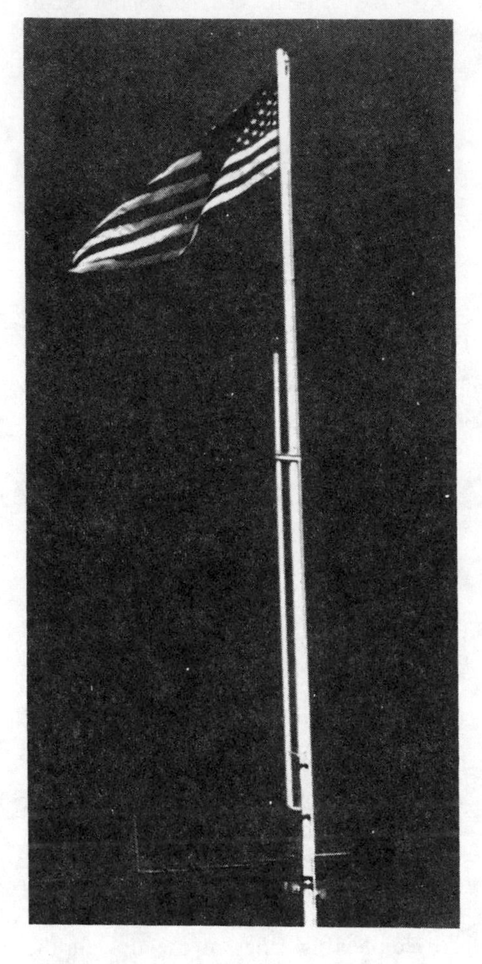

Construction

Because I live near the ocean, I decided to make the antenna from copper pipe to minimize corrosion. But there's no reason why galvanized steel pipe, or perhaps thick-walled aluminum tubing, couldn't be substituted. The Flagpole J is made with conventional copper plumbing pipe, which is readily available at hardware stores. I used 1½-inch pipe for the bottom half of the antenna, reducing it to 1¼-inch pipe for the top half. The J section is made from ¾-inch pipe.

Building the Flagpole J is simple and straightforward. Fig 1 shows how the antenna is constructed. One distinguishing feature is the length of the radiating element. In most J antennas, this element is three-quarters of a wavelength long. Because of mechanical considerations, the Flagpole J is only one-half-wavelength long.

First, cut all of the pieces of pipe to their proper lengths. Use a tubing cutter or hacksaw, and deburr the cut ends. Prepare the

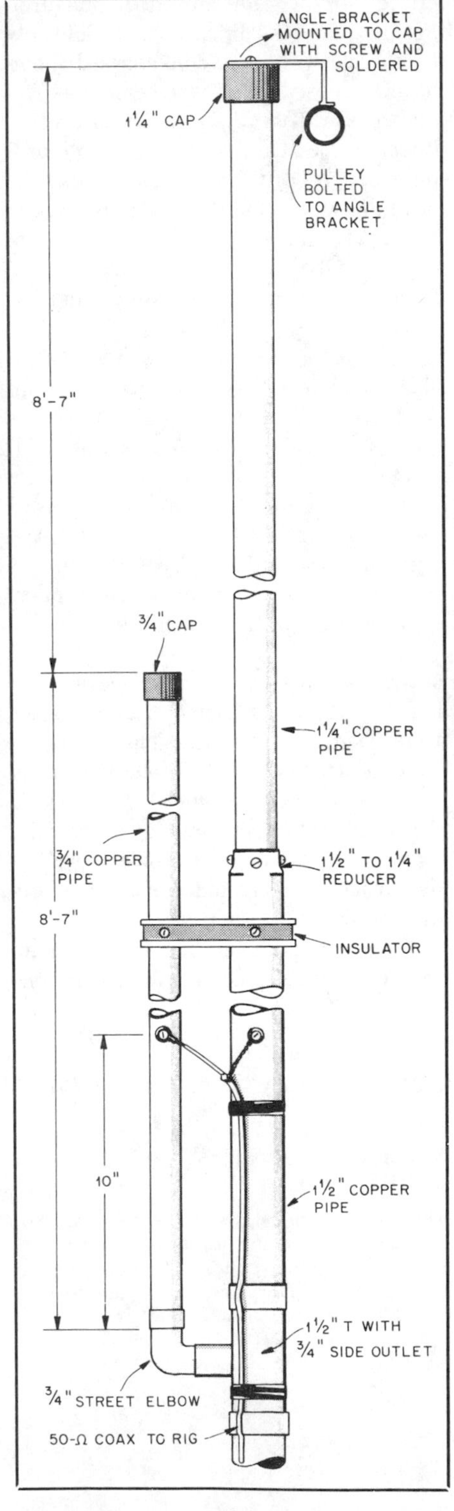

Fig 1—Construction of the 10-Meter Flagpole J.

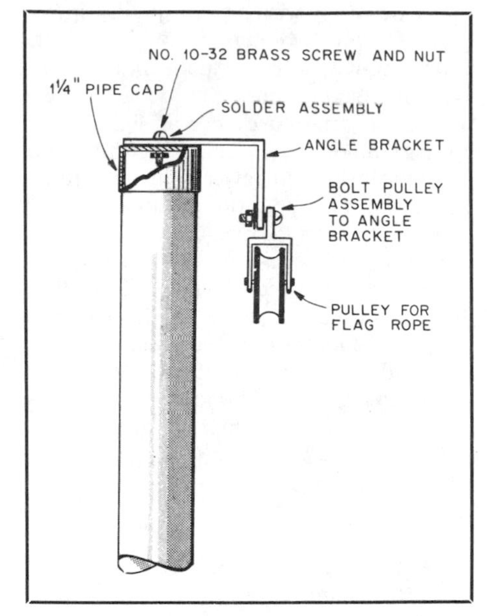

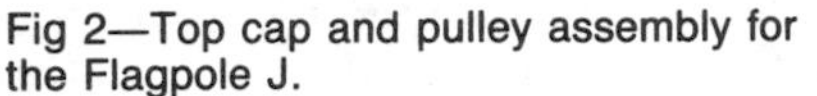
Fig 2—Top cap and pulley assembly for the Flagpole J.

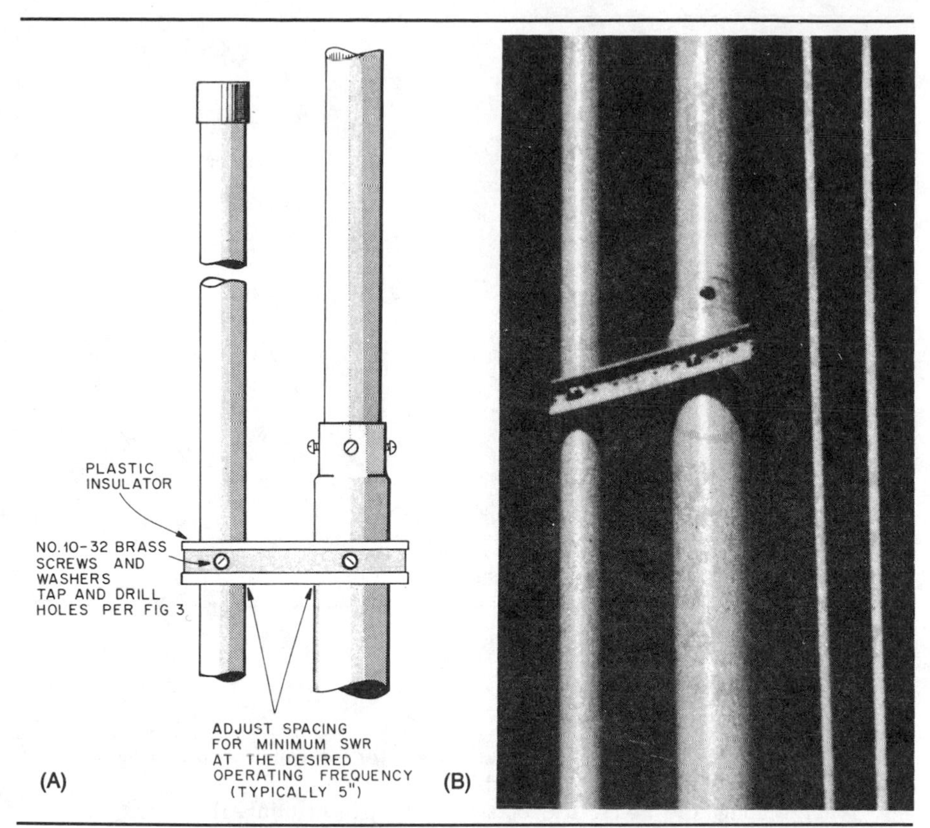

Fig 4—Detailed drawing (A) and photo (B) of the insulator attachment.

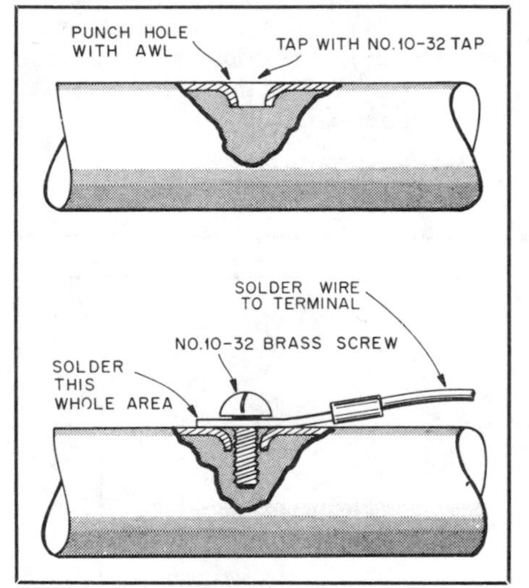

Fig 3—Feed-point attachment details.

top pulley assembly (assuming you intend to use this antenna as a flagpole, too) as shown in Fig 2. Next, solder all the pipe caps and reducers onto the various pipe sections, then finish by soldering together the antenna sections. Use soldering flux or tinner's fluid to ensure good bonding. Resin-core solder will not do the trick on copper pipe. It's best to clean the metal with flux first, then use solid-core solder to make the joint. Avoid the use of acid-core solder. Be sure to clean all the joints when you're finished soldering.

If you're going to paint the Flagpole, first perform the following steps. See Fig 3. Mark the location of the two feed-line terminals. Use a sharp punch or awl to make a hole large enough to pass a no. 10-32 tap. Punching a hole is better than

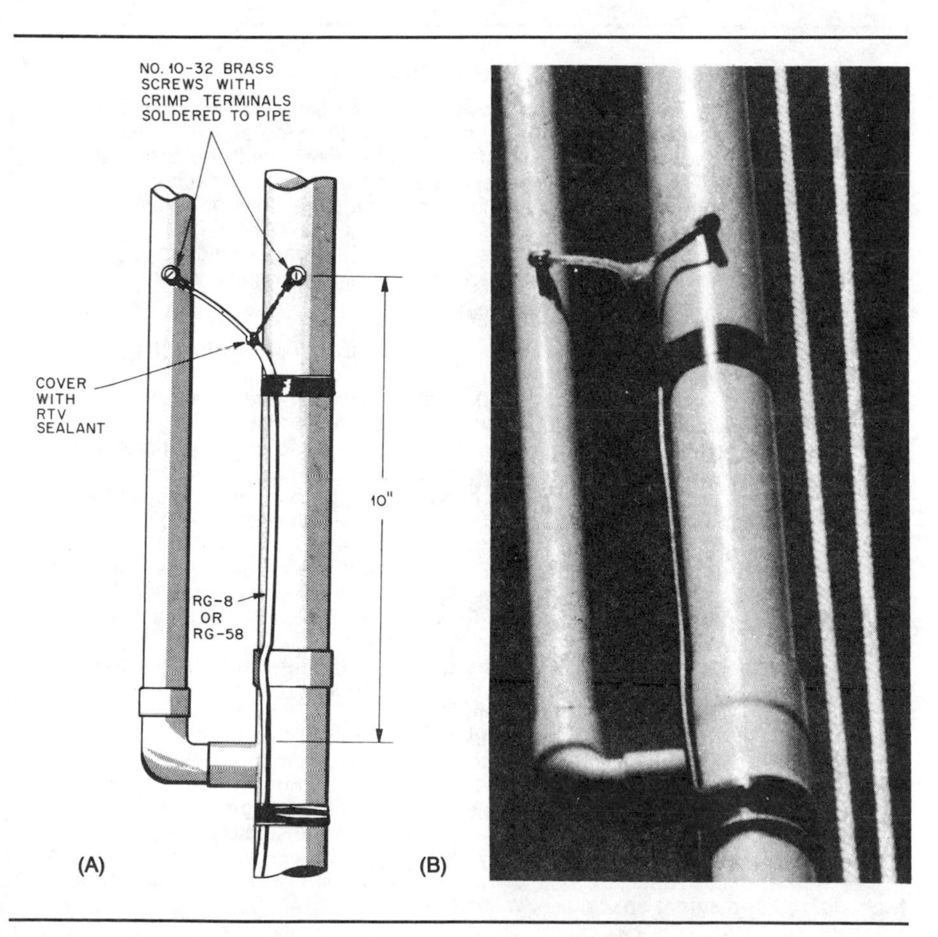

Fig 5—Feed-point attachment detail drawing (A) and photo (B). See Fig 3.

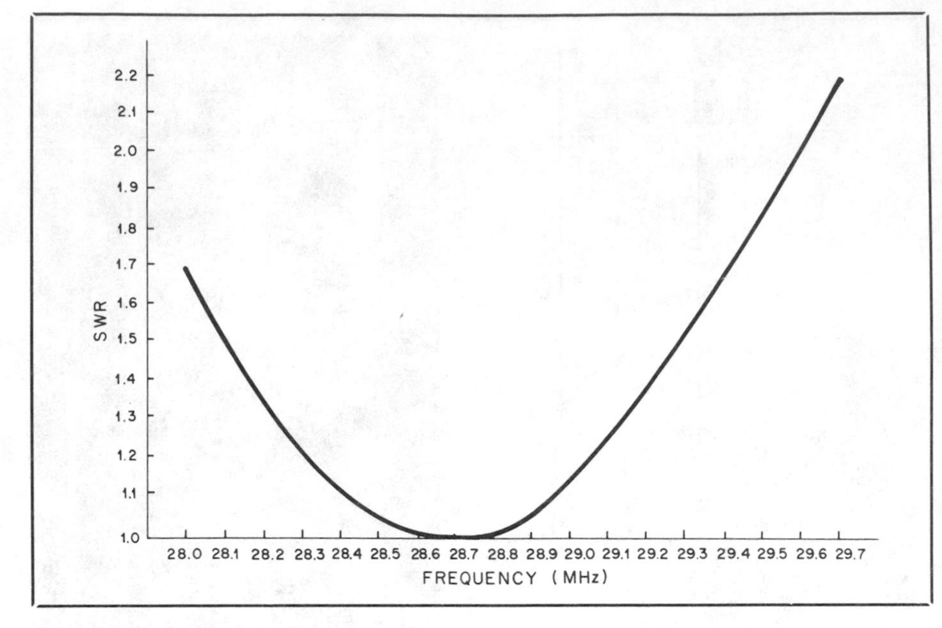

Fig 6—SWR curve of the Flagpole J.

drilling a hole because punching produces a tapered hole that is well-suited for threading. After tapping the holes, use brass screws to fasten the terminals that will connect to the feed line. Punch and tap the holes for the insulator-mounting screws as well.

Prior to painting, thoroughly clean all exposed metal parts. Mask the area around each feed-line terminal, then spray the entire assembly with a good-quality epoxy paint. If your decor allows, use white paint; it will reduce thermal expansion and contraction, making your paint job last longer. When painting, use several light coats rather than one heavy coat. The finish will be smoother, with fewer drips and runs, and it will also be more durable.

When the paint is dry, attach the plastic insulator with brass screws as shown in Fig 4. I used a piece of UV-resistant PVC for the insulator. After the insulator is attached, connect the feed line to the two terminals as shown in Fig 5. Solder the center conductor and braid to the terminals using resin-core solder. Then, apply RTV® silicone sealer to the end of the feed line to seal out moisture.

Tune-Up

Mount the antenna in an area that is free of obstacles and allows you to reach the insulator to adjust its spacing. Set your rig to the center of your desired operating range and measure the Flagpole's SWR—it should be less than 3:1 right off the bat. (If not, check for such things as shorted feed-line conductors, wrong element lengths or bad coax-connector installation.) Adjust the insulator spacing for minimum SWR. You should be able to get it very close to 1:1.

Building The Flagpole J For Other Bands

You can scale the Flagpole J up or down to operate on a variety of bands—all the way from UHF down to 160 meters. At 2 meters and above, a few constructional changes are in order: No stub insulator is required (the antenna is very rigid with the short dimensions required) and a variable capacitor is installed at the top of the short J section. I made a capacitor from brass hardware (see Fig A).

Scaling the Flagpole for other bands is simple. The main section is roughly one-half wavelength long. Its length can be calculated using the formula

Length (in feet) = 492 / frequency (MHz) (Eq 1)

The short vertical section is about a $\lambda/4$, and is therefore one-half the length of the longer section.

The feed line is tapped up the Flagpole at a distance equal to about 1 inch per meter of wavelength, so a 2-m J would have the feed line attached about 2 inches from the bottom, a 6-m J about 6 inches, and so on. Final determination of this distance is done during tuning, described later.

The distance between the two vertical members of the J should ideally be no greater than 1% of the wavelength, or about 0.4 inch per meter of wavelength. So, a 2-m J would have the elements spaced about 0.8 inch apart, a 6-m J about 2.4 inches apart, and so on. (I used a rule-of-thumb spacing of 0.5 inch per meter of wavelength with no observable problems.) The element spacing can be adjusted for minimum SWR at the top of the smaller vertical member—a sort of fine-tuning point.

Once you build the Flagpole J for the desired band, it must be tuned to match the feed-point impedance to that of the feed line. The tuning process is an interactive one with two adjustments. First is the location of the feed-line tap, second is the spacing between the two elements. To find the proper feed-line tap point, attach the center conductor and shield to the Flagpole with plastic wire ties. Then, move the feed point up and down until you find the point of minimum SWR. Now, adjust the element spacing for minimum SWR. Alternate between the two adjustments until an absolute SWR minimum is reached; it should be very close to 1:1. If you've built the Flagpole with a tuning capacitor as shown in Fig A, adjust the capacitor instead of changing the element spacing.—*Jim Hendershot, WA6VQP*

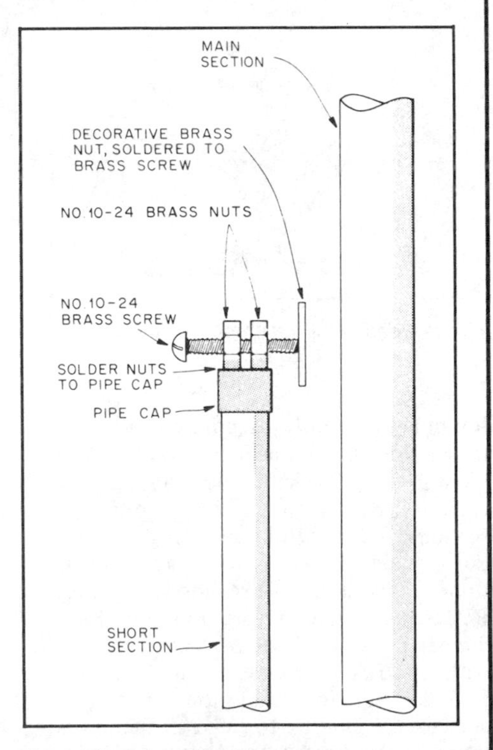

Fig A—Details of the homemade capacitor for VHF/UHF versions of the Flagpole J. The capacitor plate can be made from a piece of sheet brass. (I used a decorative brass nut that is part of a lamp-shade support.)

Once the initial tuning is done, mount the Flagpole at its permanent position. Make sure the antenna is clear of power lines. Use a solid mounting base and substantial hardware to secure the Flagpole J. Remember, this much metal has a lot of wind load, and average house eaves won't handle it well.

Summary

The SWR curve of my Flagpole is shown in Fig 6. I have my Flagpole tuned to favor the lower end of the 10-meter band, but I can QSY to the upper part of 10 meters without trouble. Results with the Flagpole J have been very gratifying. I've received excellent signal reports. I've no more "hot shack" problems: this is an elevated-feed antenna design, so it does not require radials or any other type of counterpoise. But what really makes this antenna shine is that my wife really likes her new flagpole! I was afraid she might complain about the J element. Instead, she commented on how nice it was to have a "modern flagpole" design with that extra little pole on the side. Well, whadda ya know...

from April 1985 *QST* (Technical Correspondence)

MODIFIED BUTTERNUT VERTICAL FOR 80-METER OPERATION

☐ Most amateurs know that shortened (inductively loaded) antennas exhibit a relatively narrow bandwidth. Most multiband vertical antennas have quite a bit of inductive loading, at least on the lowest one or two bands. They perform satisfactorily only over a narrow frequency range within these bands.

I became tired of readjusting my Butternut HF6V vertical antenna on 80 meters every time I switched from phone to CW operation, and vice versa. Although a Transmatch allowed me to feed the antenna over the entire 80-meter band with a low SWR at the transmitter end, I felt that the antenna was performing poorly when I was operating more than about 30 kHz from the resonant frequency. The HF6V antenna is only 26 feet long. If a shortened antenna shows a low SWR over a wide frequency range, it probably indicates that the ground system and/or feed line is so lossy that the overall radiation efficiency of the antenna system has been greatly reduced.

Unlike most multiband verticals, the Butternut HF6V antenna uses an external air-wound inductor near the bottom of the mast as an 80-meter loading coil. The lower end of the coil is electrically and mechanically fastened to the mast with an aluminum clamp that is secured with a bolt and wing nut. The antenna is tuned to resonance by loosening the wing nut and stretching or compressing the coil.

To make it easier to set up the antenna for optimum performance in either of two preselected band segments, I installed a tap at about 2½ turns from the bottom of the coil. With this tap connected to the bottom end of the coil, I could operate on the phone segment (around 3795 kHz). If the tap was left open, I could operate on the CW segment (around 3505 kHz). Although I was able to shift between band segments quickly and accurately without loosening the wing nut, I was unhappy with the need to change the tap manually.

To solve that problem, I installed a medium-power SPDT relay in a plastic refrigerator box and taped the box to the antenna mast just below the coil. The movable relay contact connects to the mast at the point where the shorting capacitor strap is clamped (see Fig. 4, point A). The normally open relay contact connects to the tap, and the normally closed contact goes to the bottom of the coil. When the relay is closed, the applied inductance is less than when the relay is released. I energize the relay to operate on the phone segment of the band.

To avoid circulating-current losses caused by shorted turns when the tap is connected, I wrapped the mast with polyethylene tape under the lower coil clamp (point B on Fig. 4). This effectively insulates the bottom end of the coil from the mast.

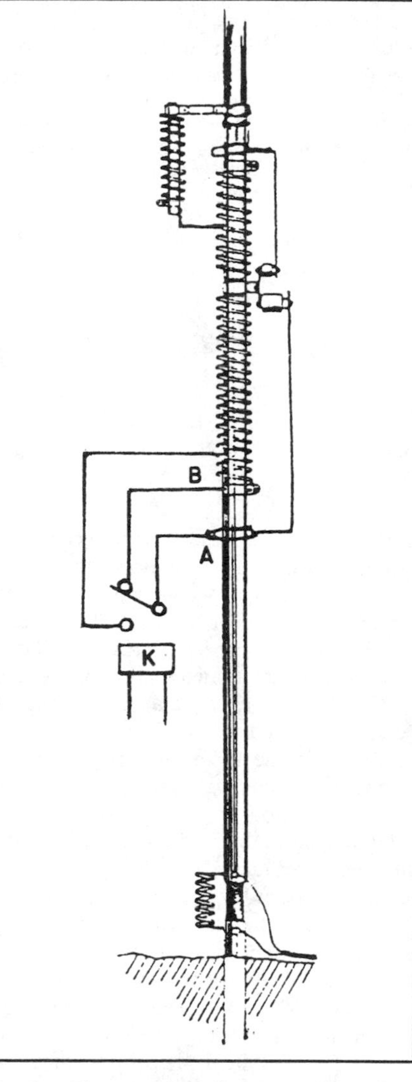

Fig. 4 — Diagram of the Butternut HF6V antenna, showing the addition of a relay and coil tap to enable operation on two segments of the 80-meter band. A is the attachment point for the movable relay contact, and B shows where the bottom coil clamp should be insulated from the mast.

After installing and connecting the relay (keep all leads as short as possible), adjust the coil length to set the desired lower operating frequency. Because of the distributed inductance and capacitance of the relay leads, this setting will be different than it was without the relay connected. Once the lower operating frequency has been established, adjust the tap position for the desired upper operating frequency. These changes did not affect the operation of my antenna on any of the other bands.

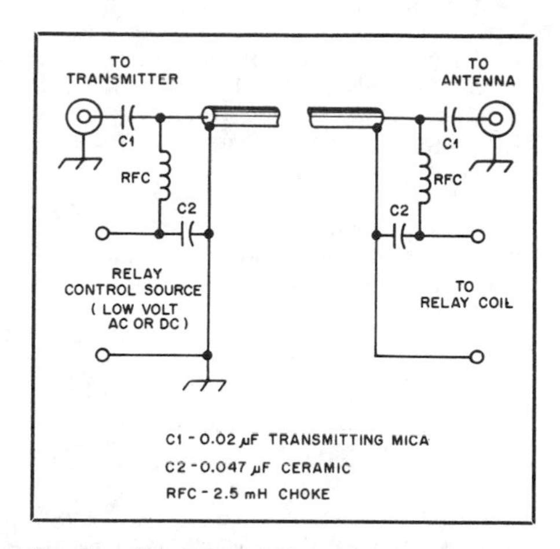

Fig. 5 — Schematic diagram showing the components required to use the coaxial-cable feed line to carry the relay-coil operating current. C1 must be a transmitting-type mica capacitor capable of withstanding the full transmitter power. C2 is a ceramic unit rated to handle the relay-coil voltage. The RF chokes must be made from wire heavy enough to carry the relay current without appreciable voltage drop.

You could run a separate control line for the relay, but I used the coaxial-cable feed line for that purpose. Fig. 5 illustrates the connection scheme. The purpose of the two capacitors labeled C1 is to keep the control voltage from traveling down the cable to the transmitter or antenna. The RF chokes prevent the radio signal from getting into the relay voltage source or relay coil, and the capacitors labeled C2 shunt any stray RF to ground. There is nothing critical about the component values, but they must exhibit appropriate reactances at RF and the relay-voltage frequency to route the signals to their proper places.

I now enjoy remote-controlled operation in either of two widely separated segments of the 80-meter band. The SWR at the center of each segment is nearly 1:1 after careful adjustment. — *Robert Snyder, KE2S, New York, New York*

from March 1983 *QST* (Hints and Kinks)

THE J² ANTENNA FOR 10 AND 24 MHz

□ This J² antenna was developed to cover the 10-MHz band, but with an eye toward future operation on the 24.89- to 24.99-MHz band. The antenna provides omnidirectional low-angle radiation with a single feed point. Fig. 1 shows the antenna dimensions.

On 10 MHz, the J² is configured as a 5/8-λ vertical, which exhibits a theoretical gain of about 3 dB compared to a 1/4-λ vertical. At 24 MHz, the J² becomes two in-phase, half-wave J antennas. The antenna base should be mounted not more than 2 feet above ground for best performance.[1] You should provide a few 1/4-λ radials for 10-MHz operation (23 feet). No radials will be required for the 24-MHz band, when that one becomes available for amateur use.

Matching to the base of the J² can be implemented with either an open-wire transmission line and matching network in the shack, or by means of an L network at the base of the antenna. The feed-point impedance will be high on both bands (>1000 ohms).

The antenna can be suspended from the side of a tower or from the limbs of a tall tree. Remember that both ends of the antenna are high-impedance points, so the rf voltage will be high. Use good insulators to support the main vertical wire.

The 1/4-λ stubs are held away from the main wire by means of homemade Plexiglas spreaders. The length is not critical up to a maximum of about 6 inches. Position the spacers about 1 foot apart along the stubs, to maintain an even spacing.

This antenna is a little short of being 1/2-wavelength long on 40 meters. By switching in some additional inductance at the base of the antenna, you should be able to use the J² on that band also. Operation as a 1/4-λ vertical for 80 meters should also be possible, but that would require a much more extensive radial system. — *Richard Schellenbach, W1JF, Reading, Massachusetts*

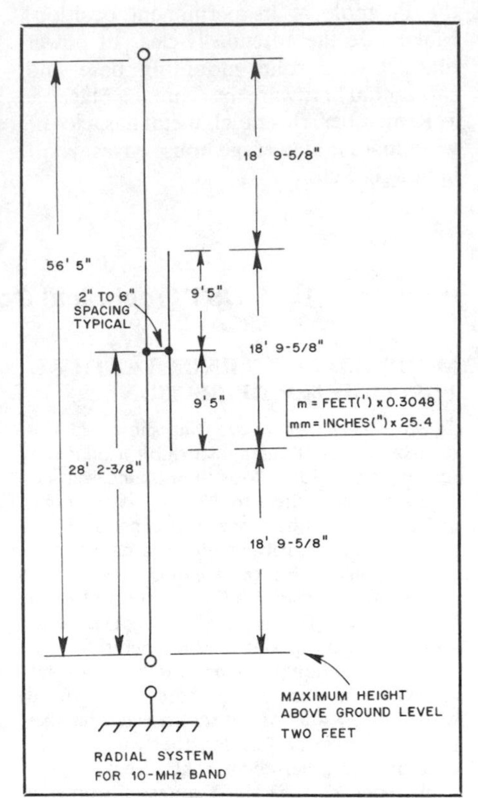

from October 1988 *QST* (Hints and Kinks)

PUTTING THE BUTTERNUT VERTICAL ANTENNA ON 160 METERS

□ My 160-meter antenna is a Butternut HF2V equipped with the optional 160-meter base coil. Considering that this antenna is only 32 feet long, it does a good job. Its bandwidth on 160 is 10 kHz or so without top loading. In fact, life on 160 with the HF2V is difficult for me only when I want to move around in the band: Readjusting the antenna is laborious.

My attempt to solve this problem was to put taps on the 160-meter base coil. I was satisfied with this until the weather turned bad. (Working at the antenna base in the freezing rain is unpleasant enough to get the old gray matter working on a better way!)

Butternut's optional 160-meter loading coil kit consists of a large inductor, two high-voltage 200-pF capacitors and mounting hardware. The two capacitors are used in parallel with the 160-meter coil to resonate the antenna in the 160-meter band. It occurred to me that easy remote tuning could be mine if I replaced one of the 200-pF fixed capacitors with a motor driven *variable* unit—to be controlled from the comfort of my shack, of course!

I calculated the inductance of the 160-meter coil as roughly 12.8 μH. Further calculations revealed that a coil of this value would require 550 pF to resonate at 1.9 MHz—more capacitance than that afforded by the paralleled 200-pF capacitors. Clearly, the antenna element adds enough inductance and capacitance to the circuit to bring the system down to its correct frequency. I decided to treat the antenna's effect as purely capacitive in calculating the value of my variable "QSY capacitor." I determined that placing a 150-to-250-pF variable in parallel with *one* of Butternut's 200-pF add-on capacitors would allow me to adjust the antenna's resonant frequency across the entire 160-meter band.

Finding a suitable variable capacitor was the next problem. I located a suitable capacitor in an electronics surplus store. Problem, though: Under test, the capacitor's 0.025-inch spacing could not withstand the voltage across the 160-meter tuned circuit. I increased its spacing to 0.05 inch by removing every other rotor plate and maintained the proper rotor-to-stator spacing by installing two rotor plates back-to-back wherever one was needed. To obtain the 150-pF minimum capacitance called for in my calculations, I rotated three of the rotor plates 180° relative to the others and paralleled the variable with a 50-pF, 3000-V mica capacitor. When three plates are fully meshed with the stator, this combination unit provides 150 pF; turning the shaft 180° meshes four plates with the stator and increases the total capacitance to 250 pF. I tested the QSY capacitor by substituting it for one of Butternut's 200-pF units. It worked great!

Next, I found a 120 V ac, 1 r/min timer motor to turn the capacitor shaft. To supply power to the motor, I decided to use two 120- to 12.6-V transformers back to back—one in the shack (step down) and the other at the antenna base (step up). The motor is coupled to the capacitor by means of a 1-inch piece of vinyl tubing cut from a fish tank suction cleaner. I mounted the QSY capacitor, motor and step-up transformer in a plastic lunch box and installed the box at the base of the antenna. To protect the components from the weather, I sealed the lunch box with caulking.

The QSY capacitor allows me to tune the antenna for a 1:1 SWR anywhere in the 160-meter band. Installation of the capacitor also had a positive effect on the SWR bandwidth at 40 and 80 meters: I gained 56 kHz at 80 and 84 kHz at 40.

Operating with the modified Butternut vertical is a pleasure. QSY on 160 is accomplished as follows. First, I determine the approximate capacitor setting by applying less than 1 W to the antenna and watching the SWR meter as the capacitor is tuned. As the capacitor approaches the correct value, the SWR drops rapidly. At this point, I turn off the capacitor motor and increase power to a watt or two. (The higher power level allows finer adjustment.) Then, I start the motor again and turn it off when the SWR reaches minimum. There you have it: 160-meter QSY in less than a minute with a few flips of a switch—from the cozy comfort of the shack!—*Robert G. Pierlott, III, WE4J, 8824 Nightingale Ln, Charlotte, NC 28226*

Chapter 4

Directional Arrays

from *The ARRL Antenna Compendium, Vol 2*

Steerable Arrays for the Low Bands

By Bob Alexander, W5AH
2720 Posey Dr
Irving, TX 75062

The use of vertical arrays on the low bands is not uncommon. Generally two elements are used, and the phase shift between elements is adjusted by means of a phasing line to provide the desired radiation pattern. Electrically steerable phased arrays that permit placement of a beam peak or null where desired are less common, and generally consist of more than two elements. A steerable, multiband phased array is rare. However, if you think in terms of time delay rather than phase shift, a two-element, multiband, beam-steerable array becomes a relatively simple project.

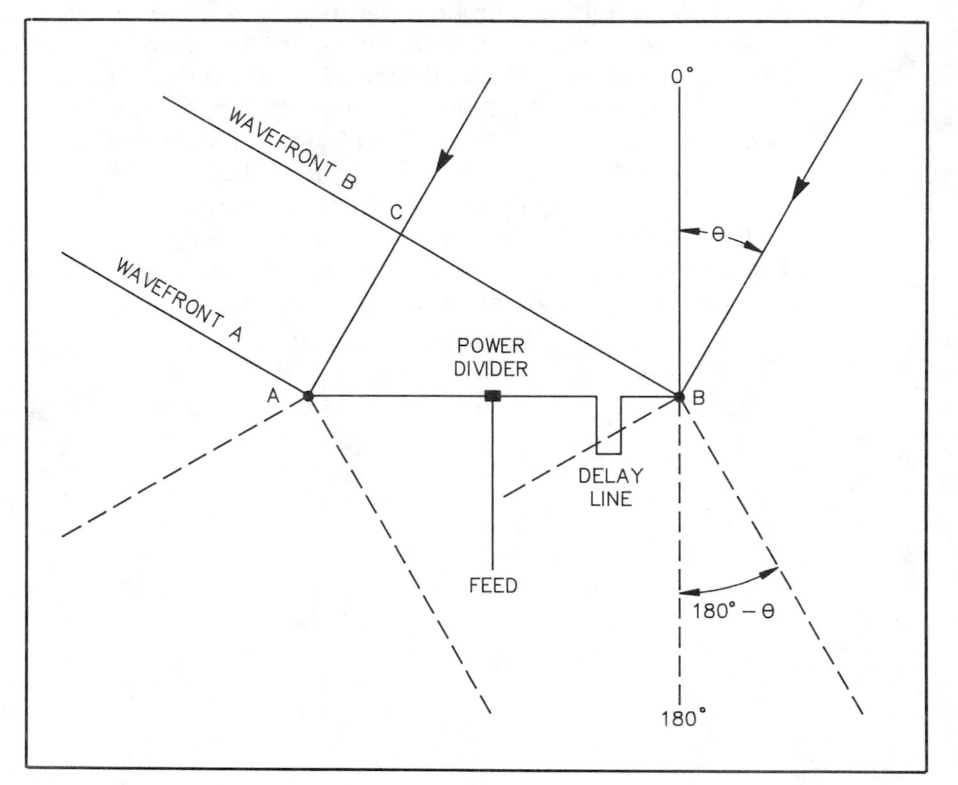

Fig 1—Array using time-delay beam steering.

Time-Delay Beam Steering

Time-delay beam steering has been previously described by Fenwick and Schell,[1] and only a simplified explanation is given here. Fig 1 illustrates two verticals with spacing AB. When elements A and B are fed with equal power through equal length transmission lines, radiation is in the broadside direction. In the case of an incoming signal arriving at angle θ, the signal will arrive at element B prior to element A. CA is the extra distance the signal must travel before arriving at element A. When a delay line electrically equal to CA is added to the transmission line to element B, the signals will add in the directions of $0° + \theta$ and $180° - \theta$. (Zero and 180° are arbitrary designations for the broadside direction.) Typical patterns for element spacings between ¼ and 2 λ are shown in Fig 2.[2]

Delay-line length is a function of element spacing and angle θ, and is given by

$\ell = VF\ S \sin\theta$

where
- VF = velocity factor of the cable in use
- S = spacing between elements (same units as ℓ)
- θ = steered angle (less than or equal to 90°)

Since the delay line, and therefore the steered angle, is independent of frequency, operation on more than one band with the same steered angle can be achieved through the use of multiband verticals and a single delay line. If provision is made to select the amount of delay and the element to which it is applied, 360° time-delay beam steering is obtained.

Circuit Description

The beam-steering system described here is based on the Omega-T 2000C Beam Steering Combiner[3], which is no longer in production. The system is rated for 1500 watts over a frequency range of 1.8 to 7.3 MHz, provided certain antenna requirements are met. These requirements are discussed later.

Equal power to the antenna elements is provided by a broadband power divider, and the amount of delay is adjusted through the use of relay switching. See Fig 3. T1 is a 2:1 transformer used to match the 25-Ω input impedance of the hybrid (T2) to 50 Ω. C1 serves to cancel the inductive reactance of the input wiring and transformer windings. C2 cancels the reactance in the relay wiring.

The hybrid difference port is terminated with a noninductive resistor, R1. When identical impedances are connected to the hybrid output ports, as in broadside operation or with ideally matched antennas, no current flows through the resistor. If a mismatch exists between the output ports, current will flow through the resistor.

Three delay lines with a combined electrical length equal to slightly less than the element spacing are switched in and out of the signal path by relays K1, K2, and K3. The element to which the delay lines are connected is selected by K4.

W4 on the schematic is 50-Ω coaxial cable. Its electrical length is equal to that of the signal path from the hybrid output through the delay-line switching relays

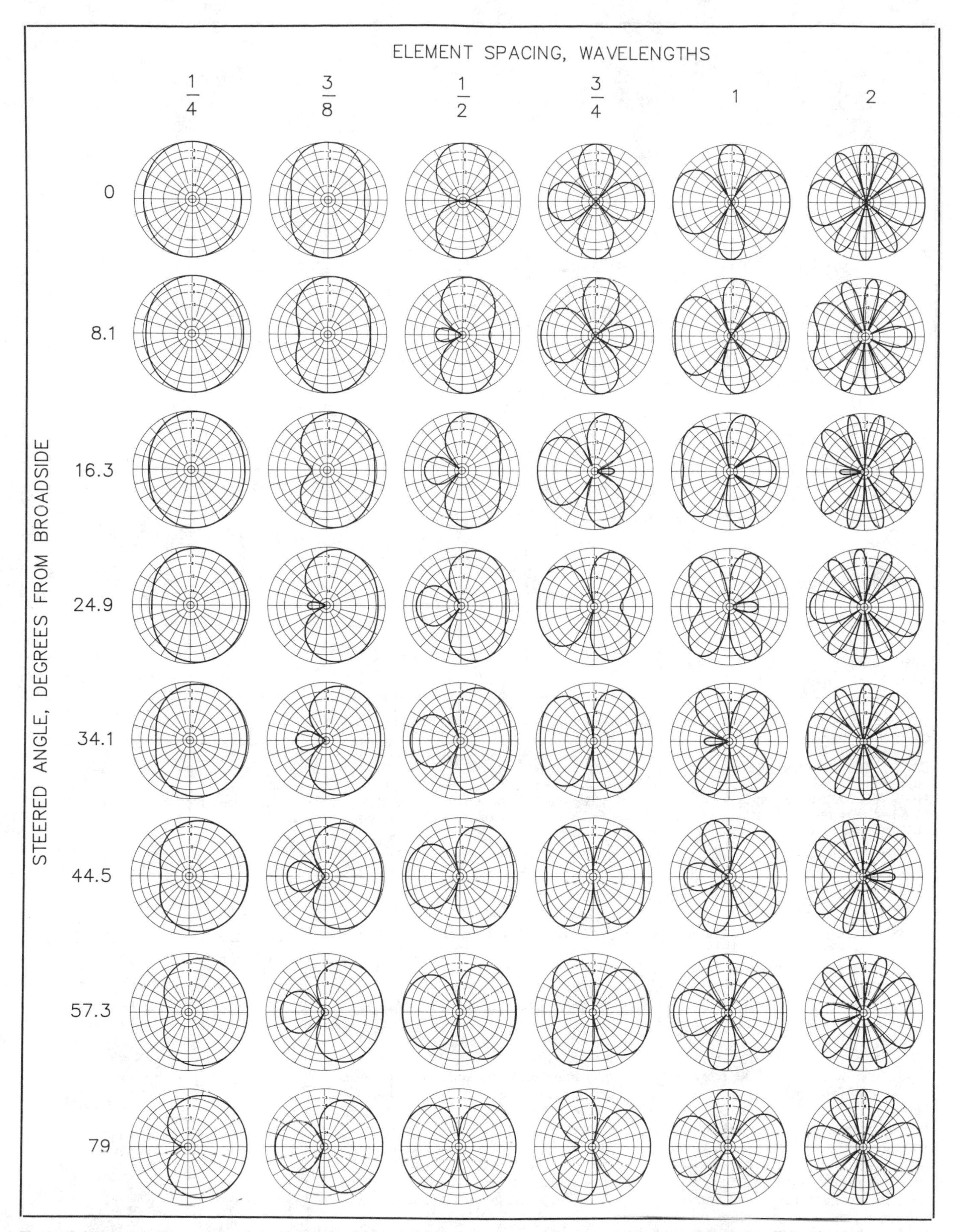

Fig 2—Azimuth radiation patterns for a time-delay beam-steering array when using element spacings of ¼ to 2 λ. The axis of the elements is along the 90-270° line. Responses are relative, shown in decibels. The array gain will vary with spacing and steering angle. Also see Table 1 and note 2.

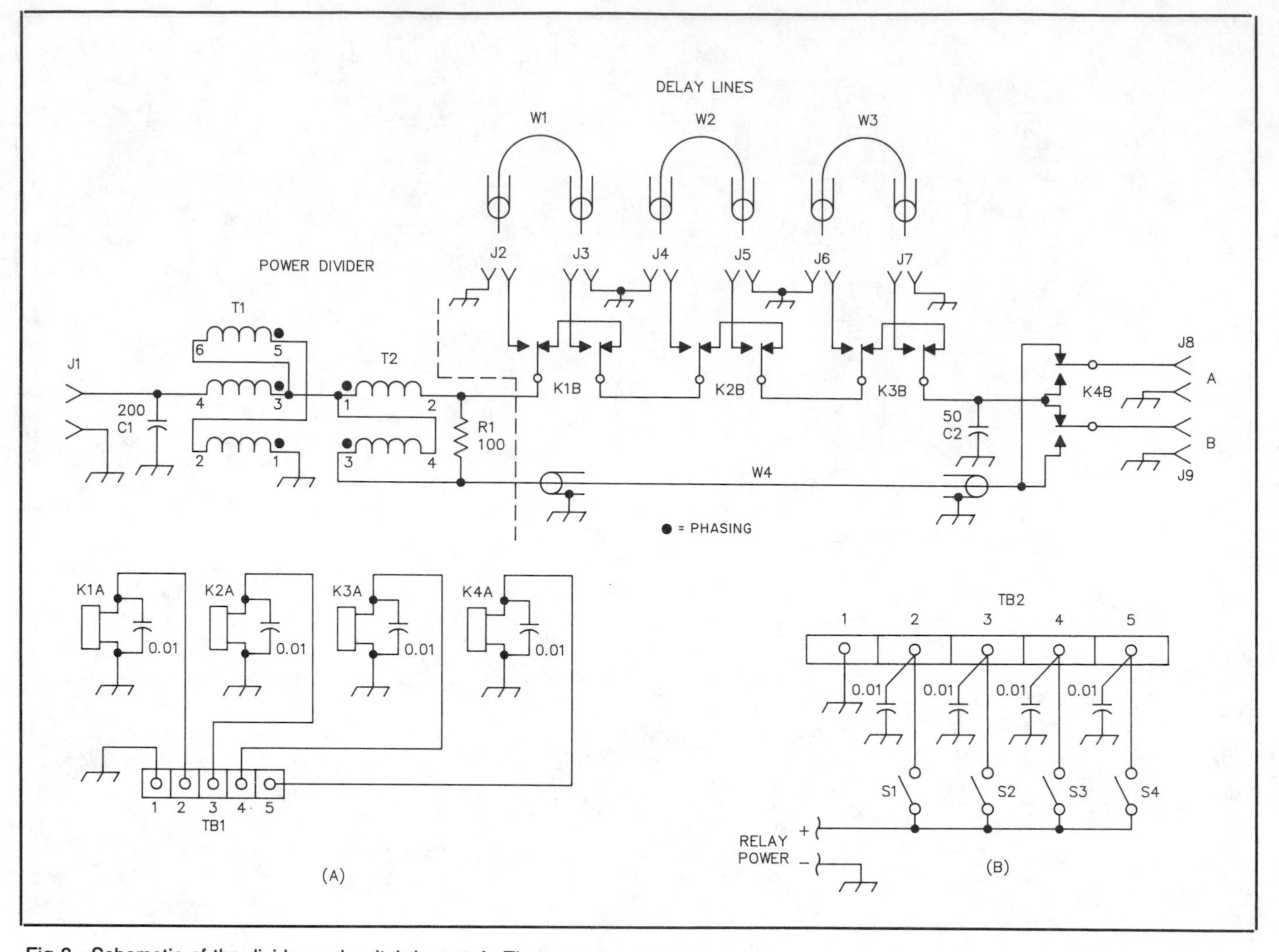

Fig 3—Schematic of the divider and switch box at A. The control unit is shown at B.

C1—200 pF transmitting-type doorknob capacitor.
C2—50 pF transmitting-type doorknob capacitor.
J1-J9—SO-239 connectors.
K1-K4—DPDT relay (contact rating and insulation must handle 750 watts).
R1—100 Ω, 100-watt noninductive resistor with mounting clips, Carborundum 886-SP101K or equiv.
S1-S4—SPST switch.
T1, T2—F240-61 toroid cores, 4 req'd (Palomar Engineers). See text and Fig 6 for winding information.
TB1, TB2—5-position terminal block.
W1-W3—50-Ω coaxial cable (RG-213 or similar; see text).
W4—50-Ω coaxial cable (RG-316 or RG-58; see text).

(relays not energized) to the antenna relay. It serves to provide an equal time delay from the two hybrid outputs to the antenna jacks. If a direct connection is made, a built-in time delay sufficient to skew the broadside pattern and the steered angles by a few degrees exists.

The control unit has been made as simple as possible, consisting of the relay power supply and four switches, as shown in Fig 3B. S1, S2 and S3 are used in a binary manner to select the desired delay line(s), as shown in Table 1. S1 selects W1 for a steered angle of 8.1° off broadside. Switching in W2 while switching out W1 steers the beam to 16.3°, and so on until all lines are selected, providing 79° beam steering. A steered angle of 90° is not needed since end-fire radiation patterns are quite broad. S4 switches the delay lines from one element to the other to reverse pattern directions.

The choice of 8.1° for the minimum steered angle is dictated by several factors. The delay-line segment lengths must be related to each other in a binary manner in order to produce smoothly increasing amounts of delay. Since the equation for delay-line length contains a sine function, then making W2 = 2 × W1 and W3 = 4 × W1 to obtain the required binary relationship results in angles greater than 2 or 4 times that chosen for determination of W1. At less than 8.1° end-fire radiation is not obtained. Therefore, the formula for determining line lengths becomes

$$W1 = VF\ S \sin 8.1$$
$$W2 = 2 \times W1$$
$$W3 = 4 \times W1$$

Table 1

Switch Position Versus Steered Angle

S4	*S3*	*S2*	*S1*	*Steered Angle*
0	0	0	0	Broadside
0	0	0	1	8.1 and 171.9
0	0	1	0	16.3 and 163.7
0	0	1	1	24.9 and 155.1
0	1	0	0	34.1 and 145.9
0	1	0	1	44.5 and 135.5
0	1	1	0	57.3 and 122.7
0	1	1	1	79 to 111 (end-fire)
1	0	0	0	Broadside
1	0	0	1	351.9 and 188.1
1	0	1	0	343.7 and 196.3
1	0	1	1	335.1 and 204.9
1	1	0	0	325.9 and 214.1
1	1	0	1	315.5 and 224.5
1	1	1	0	302.7 and 237.3
1	1	1	1	281 to 259 (end-fire)

where the W numbers = the cable length in the same units as S, VF = velocity factor and S = element spacing.

Construction

The divider and switching relays are housed in a 3 × 8 × 12-inch aluminum box. Component layout has a definite effect on the bandwidth of the power divider. The parts arrangement shown in Fig 4 should be adhered to as closely as possible.

The autotransformer and hybrid are each wound on stacked F240-61 cores prepared as shown in Fig 5. A coating of thermal grease is applied between the cores and aluminum bracket. The cores are centered on the 1.4-inch hole in the bracket and secured with tie wraps. One side of the toroids is then double wrapped with Scotch no. 61 TFE or similar tape. It should be noted that the cores will overlap the edge of the bracket by 0.2 inch on each side.

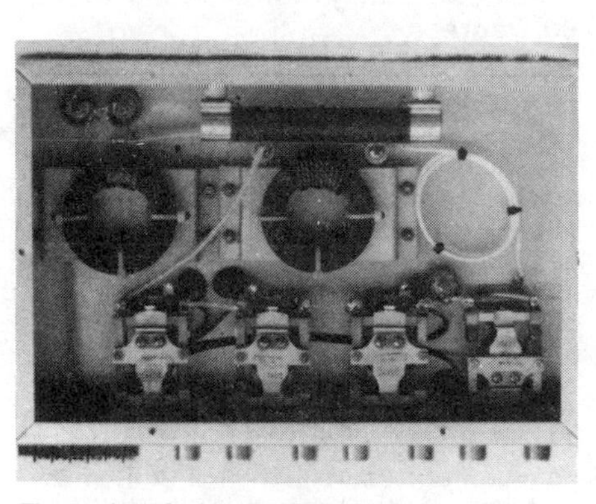

Fig 4—Interior view of the divider and switch box.

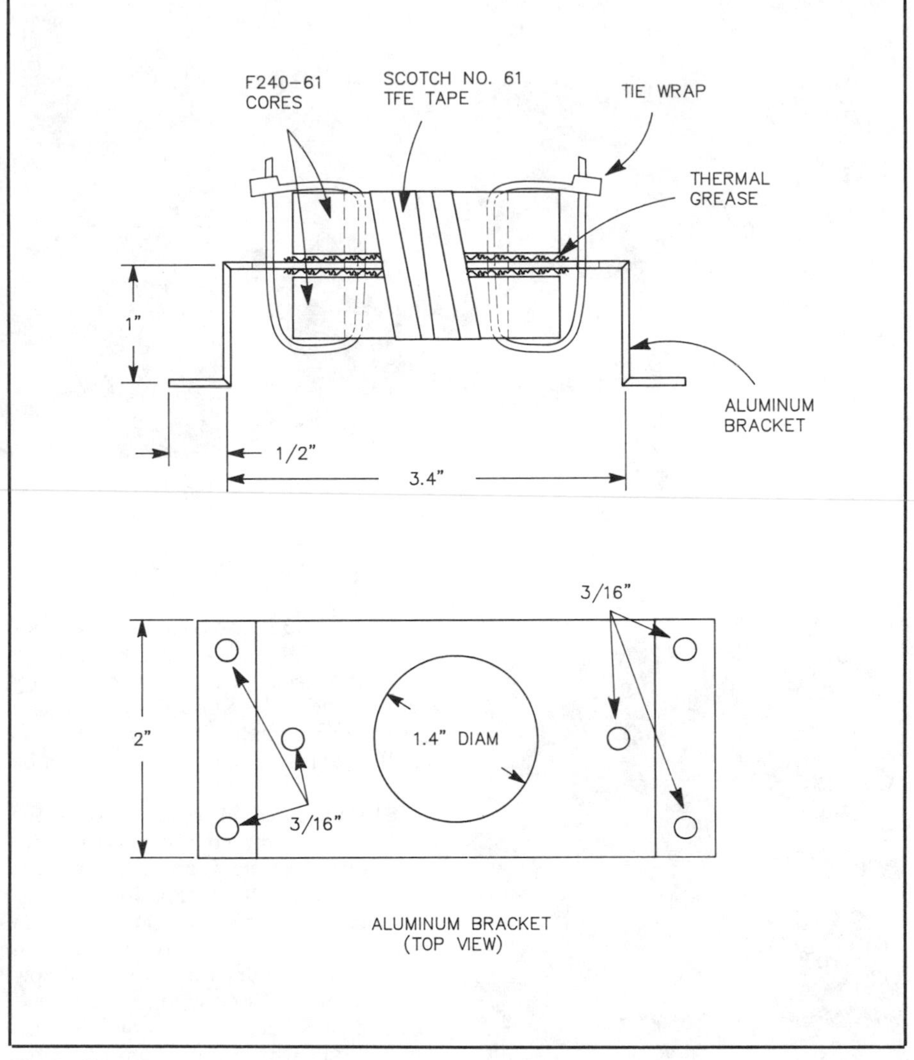

Fig 5—Transformer and hybrid assembly.

The transformer and hybrid are wound as shown in Fig 6. The transformer is trifilar wound with 5 turns in each of the outside windings and 4 turns in the center winding. The hybrid has 10 bifilar turns of twisted no. 18 enameled wires. The wires are twisted 3 times per inch prior to winding on the cores. In each case windings should be tight around the cores and bracket, and as close as possible to adjacent turns.

Transmitting-type doorknob capacitors are used at C1 and C2. Two such capacitors were connected in parallel at C1 to obtain the required capacitance. C1 should be mounted as close as possible to T1. Connections are made via solder lugs secured to the tops of the capacitors. Braid (from RG-58) with Teflon sleeving over it is used for the connection between J1 and T1 C1. Likewise, braid is used between T1 and T2.

Two ceramic standoffs, mounted adjacent to T2, are used for the hybrid output connections. The mounting clips for R1 are affixed to ceramic standoffs secured to the

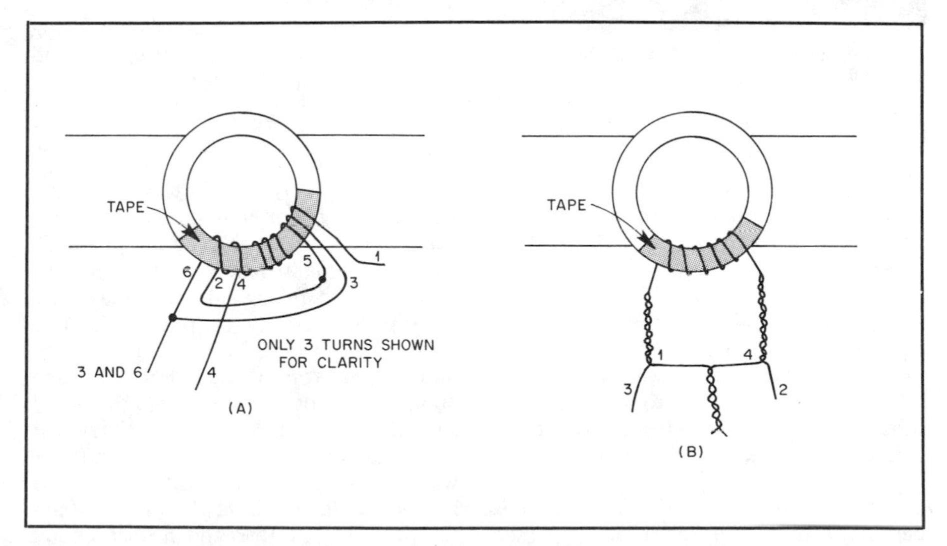

Fig 6—2:1 transformer and hybrid winding detail. Core and brackets are assembled as shown in Fig 5. The transformer (A) is 5 trifilar turns of no. 18 enameled wire, with 1 turn removed from lead no. 4. Lead 1 is ground, lead 4 is the 50-Ω input, and leads 3 and 6 are the 25-Ω output. The hybrid (B) consists of 10 bifilar turns (twisted 3 times per inch). See text and Fig 3.

Fig 7—Power divider and switch box (left) and control box (right).

Fig 9—The array installation for 80- and 40-meter operation using Butternut HF2V verticals.

Fig 8—Power divider and switch box with all cables connected.

side of the box opposite the relays. They should be positioned so the leads to T2 are short and of equal length.

The relays are mounted as close as possible to the coaxial connectors, as the connections between them must be short and direct. C2 is mounted between K3 and K4. The connection from T2 to K1 is made by using braid with Teflon sleeving over it. All connections between relays are made using braid.

W4 is 50-Ω coax cable. Its length is determined by measuring the length of the leads used from T2 to K1 to K2 to K3 to K4. The length of the signal path through K1, K2, and K3 is also added in (relays not energized). Cable length is the measured length of the signal path times the velocity factor of the cable. RG-316 was used here because of its small size and high power-handling ability. However, RG-58 can be used.

TB1 is mounted on the same side of the divider box as the coax connectors. All connections between it and the relays are bypassed with 0.01-μF capacitors.

The relay power supply is contained in a separate control unit. It is not shown on the schematic nor described here, as supply voltage is dependent upon the relays used. S1 through S4 are mounted on the front panel, along with a power on-off switch (not shown on the schematic). A five-lug terminal block is placed on the rear of the box and each terminal is bypassed with a 0.01-μF capacitor.

The assembled boxes are shown in Fig 7. Fig 8 shows how the power divider and switching box is installed. The complete array appears in Fig 9.

Testing

After assembly is completed and all wiring checked for errors, the input SWR should be measured. Terminate J8 and J9 with 50-Ω loads (each load should be capable of dissipating half of the maximum input power). Apply low power at J1 and measure the input SWR at the bottom, middle and top of the 160-, 80-, and 40-meter bands. In all cases, the SWR should be less than 1.5 to 1. Worst-case SWR measured on the unit described here was 1.20 at 7.3 MHz. An SWR of 3 or higher results if either J8 or J9 is not terminated in 50 Ω, or if an open circuit condition exists between J8, J9 and the hybrid. If all appears normal after the

Delay-line Design for Steering to 90 Degrees

Author Bob Alexander, W5AH, presents an excellent idea for a directional array that uses two vertical antennas. Bob chose to steer from 0° (broadside) to 79°. How would you go about modifying his design if you wished to steer to 90°?

The answer is to change the length of the delay lines so that their total electrical length is equal to the physical spacing between the antennas. To do this you will need three binary weighted lengths (as Alexander has pointed out). You could think of them as binary digits. The sum of the digits is seven (1 + 2 + 4 = 7). That means that the delay-line electrical lengths are 1/7, 2/7 and 4/7 of the physical spacing between elements. The delay/spacing ratio is the sine of the arrival (or steering) angle, θ. The following table shows the relationships for a system designed to steer to 90°.

Fraction	*(Sine)*	*Angle, Degrees*
0/7	0.0000	0.00 (Broadside)
1/7*	0.1429	8.21
2/7*	0.2857	16.60
3/7	0.4286	25.38
4/7*	0.5714	34.85
5/7	0.7143	45.58
6/7	0.8571	59.00
7/7	1.0000	90.00 (End fire)

* = delay-line lengths

To determine actual line lengths, multiply the value in either the fraction or sine column (they're the same) times the physical spacing between elements times the velocity factor of the coaxial cable used for the delay lines.—*Chuck Hutchinson, K8CH*

tial measurements, repeat them using increased power. There should be no change in SWR.

Next, connect the control unit to the divider and switch box with the 5-conductor cable to be used in the final installation. Apply power and visually check that the correct relays are energized in each of the switch positions.

Antenna Requirements

The two antenna elements must have the same SWR with respect to each other at all frequencies of operation. In other words, the SWR curves should be identical. The impedance of each antenna with the other antenna removed should be as close to 50 Ω at resonance as is practicable.

They should not be spaced more than 2 λ at the highest operating frequency nor closer than ¼ λ at the lowest frequency. For 160- through 40-meter operation, the elements should be spaced between ¼ and ½ λ on 160 meters (1 to 2 λ on 40 meters). For best performance, the antennas should not be near any surrounding metal objects.

Installation

Once the antenna spacing has been selected, the three delay lines are fabricated from low-loss 50-Ω cable using the formulas given earlier. Length is measured from tip to tip of the coax connector center conductors, and must be as near as possible to the calculated values.

The divider and switching box is placed midway between the antennas. Feed lines from the box to the antennas can be any length so long as they are of equal length. Hookup is as shown in Fig 3.

Operation

The SWR seen by the transmitter will vary somewhat as the steered angle is changed. Although the SWR seen at the transmitter may be quite low, operation on frequencies where the antenna SWR exceeds 2:1 should be avoided, as damage to the divider can result. The impedance seen at the transmitter and that seen by the divider is the same only for broadside operation.

On receive, the pattern peaks and nulls will be most noticeable on signals arriving at relatively low angles. With signals arriving at high angles, the steered angle selected for best reception may not accurately indicate the azimuth of the transmitting station.

Table 1 lists the switch positions and corresponding steered angles. The increasing rate of change in the steered angle is compensated for by the fact that the beamwidth increases as the steered angle nears the end-fire direction.

Summary Comments

1) Element spacing need not be exactly ¼ or ½ λ, or any of the values listed in Fig 2. An element spacing of, for example, 0.32 or 0.86 λ will work just as well.

2) All coaxial cables, especially those used for the delay lines, should be of the low-loss variety. Lossy cables will hurt performance, especially where null depths are concerned. A loss of 0.8 dB in the delay lines will limit nulls to 20 dB or less.

3) Never change beam headings while transmitting, as the relays may be damaged.

4) Since the divider and switch-box assembly is to be mounted outdoors, the box and all connectors should be weatherproofed.

5) The beam-steering method described here is not new. It has, however, seen little use. Hopefully, the material presented here will encourage the use of this older technology.

I wish to express my thanks to Dick Fenwick, K5RR, for his assistance in preparing this article.

Notes

[1]R. C. Fenwick and R. R. Schell, "Broadband, Steerable Phased Arrays," *QST*, Apr 1977.

[2][The theoretical patterns of Fig 2 are based on no losses and equal *currents* in the elements, rather than equal power. Because of mutual coupling, element feed-point impedances will not be the same except at a steering angle of 0°, and equal power will therefore yield unequal currents. Further, the phase relationship of the currents may not be as intended. As a result, the lobes may be skewed slightly, and the deep pattern nulls shown in Fig 2 may not be realized in practice. Detailed information on this phenomenon appears in Chap 8 of the 15th edition of *The ARRL Antenna Book* (1988).—*Ed.*]

[3]The Omega-T 2000C Beam Steering Combiner is a patented product of Electrospace Systems Inc, Richardson, TX.

Phased Verticals with Continuous Phase Control

By Peter H. Anderson, KZ3K
Department of Electrical Engineering
Morgan State University
Cold Spring Lane & Hillen Rd
Baltimore MD 21239

This paper describes a simple arrangement for phasing two verticals. The arrangement utilizes a simple symmetrical pi network to provide a continuous phase delay and avoid the inconvenience of switching transmission-line sections.

The continuous phasing arrangement may be used on several bands. In fact, I have used this circuit to great advantage on 80, 40 and 20 meters. By changing the angle, or "steering" the array, I have been able to consistently vary signals by as much as 20 dB. The required parts are inexpensive and readily available. Construction is straightforward and simple.

I have two Butternut verticals spaced about 60 feet apart. They were erected without plans for future phasing and were tuned to different frequencies on the 80-meter band. The possibility of phasing them was posed to my undergraduate electrical engineering design class. They seized upon the idea and soon developed a workable system. When we put the circuitry on the air, we were pleasantly surprised with the results.

Most phasing arrangements are presented on the premise that the user can feed the antennas in phase. In my installation, the relative lengths of the coaxial runs is unknown. In addition, one of the verticals has a 160-meter coil and the other has a 30-meter coil. As a result, I have no idea of the relative electrical lengths from the shack to the antennas, which of course will vary from band to band. Under the circumstances it seemed as though the benefits of phasing were clearly reserved for amateurs who possessed two identical antennas, lots of coaxial cable, and a great deal of leisure time.

Fortunately for the rest of us, this reasoning is flawed. By adding a 0° to 90° variable phase lag to either of the two antennas, and by reversing the phase as necessary, it is possible to adjust the relative phase of one antenna to the other by a

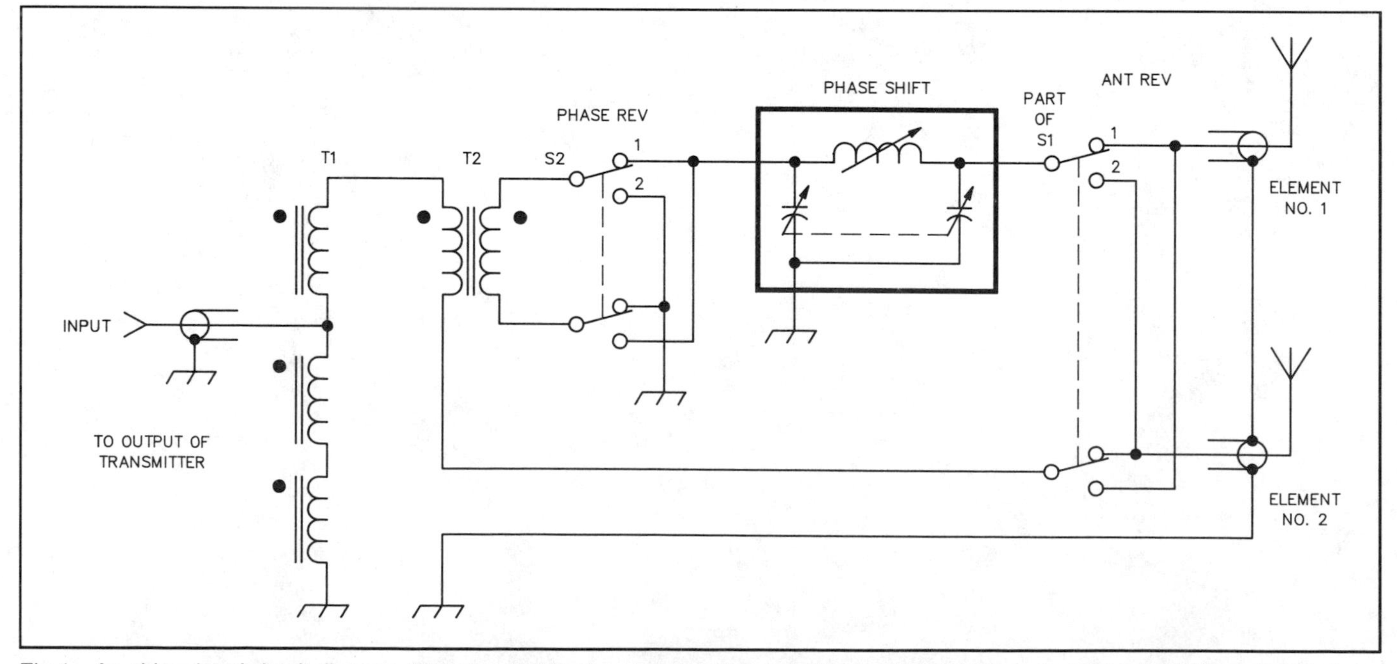

Fig 1—An abbreviated circuit diagram. The phase of the input to the phase shift network may be inverted using DPDT toggle switch S2. S1, used to interchange the antennas, is part of a larger function switch discussed in the text.

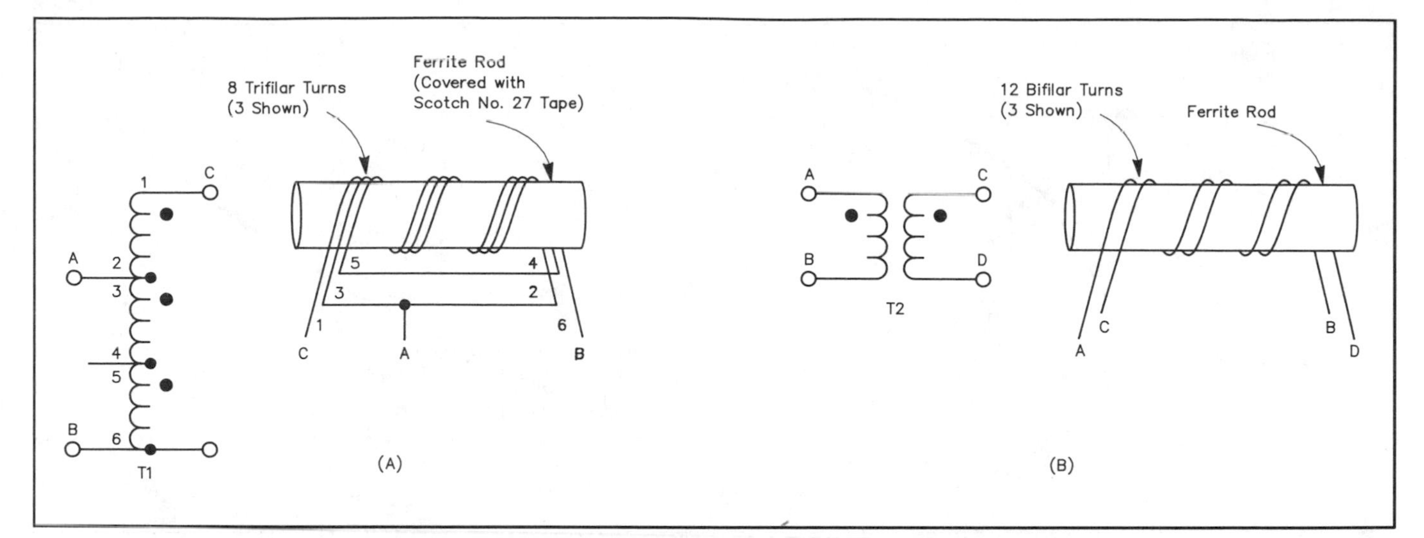

Fig 2—Details of transformers T1 and T2. Both transformers use a 7½-inch ferrite rod (Amidon R61-050-750). The rods are first taped using Scotch no. 27 glass cloth electrical tape. The no. 14 AWG thermaleze wire is available from Amidon. T1, shown at A, consists of eight trifilar turns connected as illustrated. The three conductors should be tightly bound together, either by twisting or by using small pieces of tape. (The length of each conductor is about 18 inches.) To prevent the rod from moving within the coil, small pieces of tape or epoxy may be used. T2, at B, consists of 12 bifilar turns which are similarly wound on a ferrite rod. The length of each conductor is about 24 inches. Amidon offers a special kit consisting of the two rods, 10 feet of wire and a 66-foot roll of Scotch no. 27 tape. When contacting Amidon, refer to this article by title and author. (See note 5.)

full 360°. The fact that the actual phase lag is unknown is unimportant. With the system described in this article, the user will be free to vary the phase so as to effectively feed the antennas broadside, end fed toward element no. 1, or end fed toward element no. 2. By using the phase circuitry to "peak" the antenna system for maximum received signal strength, the user will optimize the transmitted signal pattern as well.

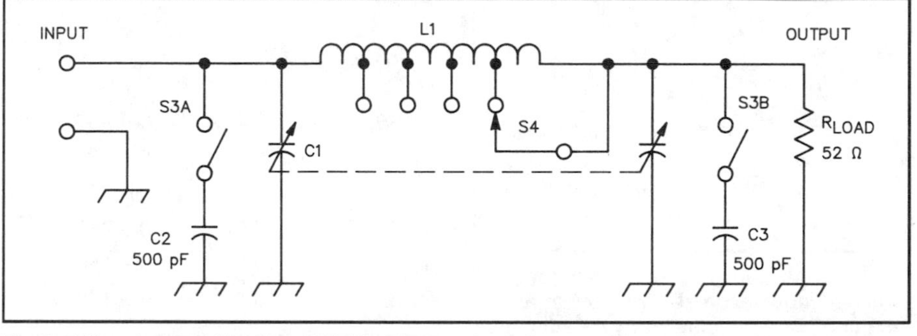

Fig 3—The phase-shift network. This symmetrical pi network may be used to add phase delay while maintaining an input impedance that is close to the 52-Ω load resistance. Using S3, C2 and C3 may be switched in to provide a 0- to 1000-pF range for use on 80 meters.

C1—Dual ganged 0-500 pF air variable capacitor.

C2, C3—500-pF doorknob type capacitor.

L1—See Fig 6 for construction details.

S3—DPST toggle.

S4—Single pole rotary switch with multiple positions to switch in the desired inductance.

Overall Configuration

An abbreviated circuit diagram is illustrated in Fig 1. Transformer T2 "floats" one of the antennas, permitting the antennas to be fed in series. Therefore, assuming that both antennas are at 52 Ω, the impedance of the series pair is 104 Ω. A simple 2:3 turns-ratio transformer (T1) transforms this impedance to 4/9 × 104 or 46 Ω. Use of these broadband transformers permits the same circuitry to be used on many different frequencies. The construction details of T1 and T2 are presented in Fig 2.

A symmetrical pi network of the type illustrated in Fig 3 is used to provide a phase shift while maintaining the 52-Ω input impedance. Note that the capacitors on either side of the inductor are equal.

Fig 4 may be used to determine the values of C (on each side of the inductor) and L to provide any phase in the range of 0° to 90°. For example, to obtain a phase shift of 45° at 7.1 MHz, the L and C values are determined to be slightly more than 0.8 μH and 175 pF. Note that Fig 4 was developed for a frequency of 7.1 MHz and it is reasonably accurate for the entire 40-meter band. However, you can use Fig 4 for any frequency by simply scaling the values.

For example, assume that the frequency is 3.905 MHz and 45° of phase lag is desired.

$L = 7.1/3.9 \times 0.8 = 1.45\ \mu H$

$C = 7.1/3.9 \times 175 = 320\ pF$

At f = 14.2 MHz, you would simply halve the values at 7.1 MHz; L = 0.4 μH and C = 87 pF.

Note that you could construct such a network for use on 40 meters by using a roller inductor, variable between 0 and 1.18 μH, and a dual-ganged capacitor having a range of 0 to 425 pF. With the appropriate settings, any phase in the range of 0 to 90° could be obtained. By scaling the values of L and C as shown in Table 1, the network could be used on any amateur band.

I started the project with a roller inductor and a dual-ganged 0-500 pF variable capacitor (with the capability of switching in an extra 500 pF for 0-1000 pF on 80 meters). However, roller inductors are a considerable expense for many amateurs. To make matters worse, I found the whole

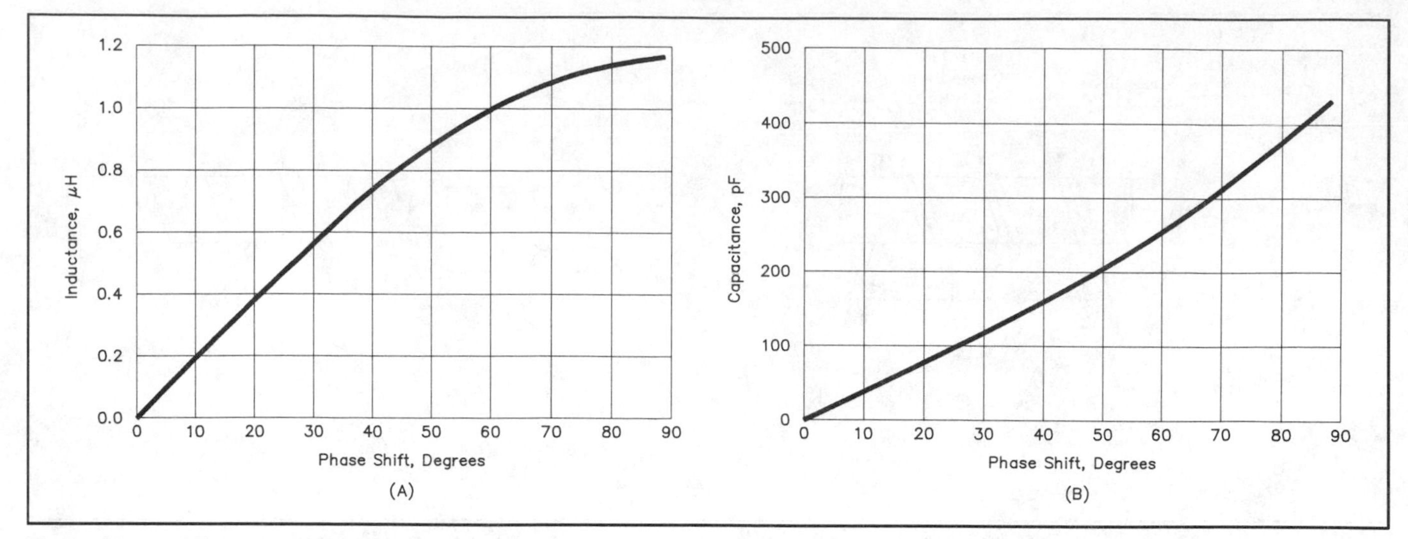

Fig 4—Phase shift versus inductance and capacitance at 7.1 MHz, for a load of 52 Ω. These plots may be used to determine the values of L and C in a symmetrical pi network to provide any phase delay in the range of 0 to 90°. The L and C values may be scaled for frequencies other than 7.1 MHz. (See text.)

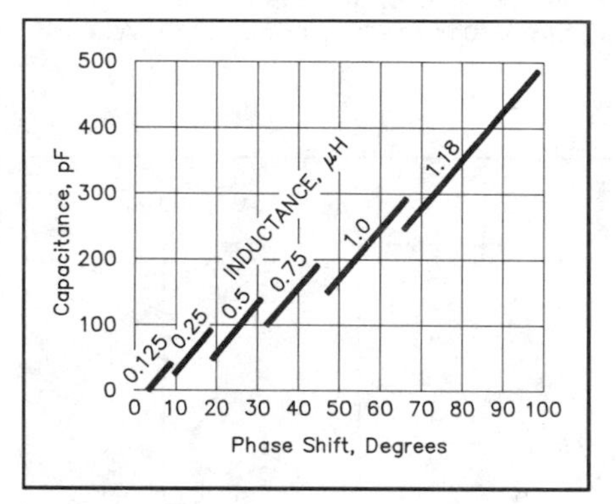

Fig 5—Capacitance versus phase for switched inductance at 7.1 MHz, for a load of 52 Ω. The phase may be adjusted by varying capacitor C over a limited range with an SWR of better than 1.2:1. For example, with an L of 1.18 µH, the phase delay may be varied from 65 to 98° by adjusting C in the range of 250 to 500 pF.

Table 1
L and C Network Values for Various Amateur Bands

Freq (MHz)	*L (µH)*	*C (pF)*
3.55	0 – 2.4	0 – 950
7.1	0 – 1.2	0 – 425
10.1	0 – 0.85	0 – 300
14.2	0 – 0.6	0 – 210
18.1	0 – 0.5	0 – 166
21.3	0 – 0.4	0 – 140

Table 2
Phase Shifts at 7.1 MHz with Fixed Inductance

Fixed L (µH)	*C range (pF)*	*Phase Range (degrees)*
0.12	0 – 50	3.1 – 9.5
0.25	25 – 100	9.5 – 19
0.50	50 – 150	19 – 32
0.75	100 – 200	32 – 46
1.0	150 – 300	46 – 65
1.18	250 – 500	65 – 98

Table 3
Phase Shifts at 3.5 MHz with Fixed Inductance

Fixed L (µH)	*C range (pF)*	*Phase Range (degrees)*
0.25	0 – 100	3.1 – 9.5
0.50	50 – 200	9.5 – 19
0.75	100 – 300	19 – 32
1.0	200 – 400	32 – 46
1.18	250 – 600	46 – 65
2.36	500 – 1000	65 – 98

Table 4
Phase Shifts at 14.2 MHz with Fixed Inductance

Fixed L (µH)	*C range (pF)*	*Phase Range (degrees)*
0.06	0 – 25	3.1 – 9.5
0.12	12 – 50	9.5 – 19
0.25	25 – 75	19 – 32
0.37	50 – 100	32 – 46
0.50	75 – 150	46 – 65
0.60	125 – 250	65 – 98

arrangement to be too confusing to use on all bands.

With this deficiency in mind, I began to explore a switched-inductor arrangement. I found that with a fixed L, the C could be varied within a limited range, achieving a variable phase while maintaining an SWR of less than 1.2. This is summarized in Table 2 and is presented graphically in Fig 5. Note that all values are at 7.1 MHz.

The results are significant. By using a small switched inductor, a dual-ganged 0-500 pF variable capacitor and a very simple six-position switch, any phase in the range of 0 to 90° can be obtained. There are no cumbersome transmission-line sections to insert! Further, the pi network values can be easily scaled to provide similar phase shifts on either 80 or 20 meters (see Tables 3 and 4).

You could develop a switched inductor network for any or all of the three bands. One simple technique I used was to develop a single tapped inductor (actually three inductors in series) as illustrated in Fig 6. This arrangement provides taps at 0.062, 0.125, 0.25, 0.37, 0.5, 0.6, 0.75, 1.0, 1.18 and 2.36 µH, and may be used on all three bands. (If you desire three separate tapped coils, one for each band, the construction data can be easily adapted.)

Operation

Refer to the detailed schematic shown in Fig 7. S1 (multiposition) permits the user to feed *only* element no. 1 or no. 2. These switch positions also allow the user to assure that both antennas present the same impedance to the phasing network. Two other positions on S1 insert the phase-shift network in element no. 1 or no. 2. These are the primary operating positions and are labeled PHASE 1 and PHASE 2. An additional position (TUNE) is provided to connect the input (through the phasing network) to a dummy load. This is especially helpful when tuning the phasing network. Another position connects the input directly to a dummy load.

After the impedance to both antennas has been verified to be roughly equal and the transmitter has been loaded using the dummy load, the phasing network can be adjusted. With S1 in the TUNE position, a particular L value is selected and C is adjusted for minimum SWR. By repeating this procedure at various L values, you can quickly build a table of the C values corresponding to the L setting for each band. You will also note that there is a range of

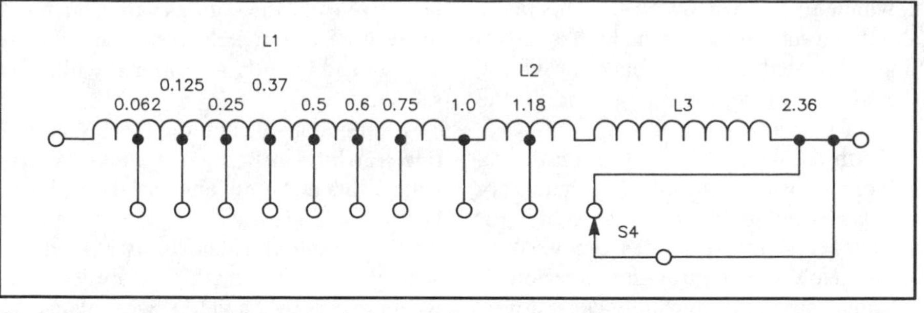

Fig 6—Construction of the tapped phase shift inductor. L1 is tapped at 1.5 turns (0.062 μH), 2.25 turns (0.125 μH), 3.75 turns (0.25 μH), 5 turns (0.37 μH), 6.5 turns (0.5 μH), 7.5 turns (0.6 μH) and 9 turns (0.75 μH). The 1.0 μH tap is at the junction of L1 and L2. L2 is tapped at 4.25 turns (1.18 μH). The combination of L1, L2 and L3 provides the 2.36 μH tap. Note that in using a single switched inductor, the lower five settings are used for 20 meters. All but the highest tap may be used on 40 meters and all settings may be used on 80 meters. Some of the phase settings will overlap on 40 and 80 meters.

L1, L3—Barker & Williamson 804T, 1 inch diameter, 4 turns per inch, length 3 inches. (See note 6.)

L2—Barker & Williamson 604T, ¾ inch diameter, 4 turns per inch, length 2 inches.

S4—Centralab PA-2001, 1 pole, 2 to 12 position rotary switch (Newark). (See note 7.)

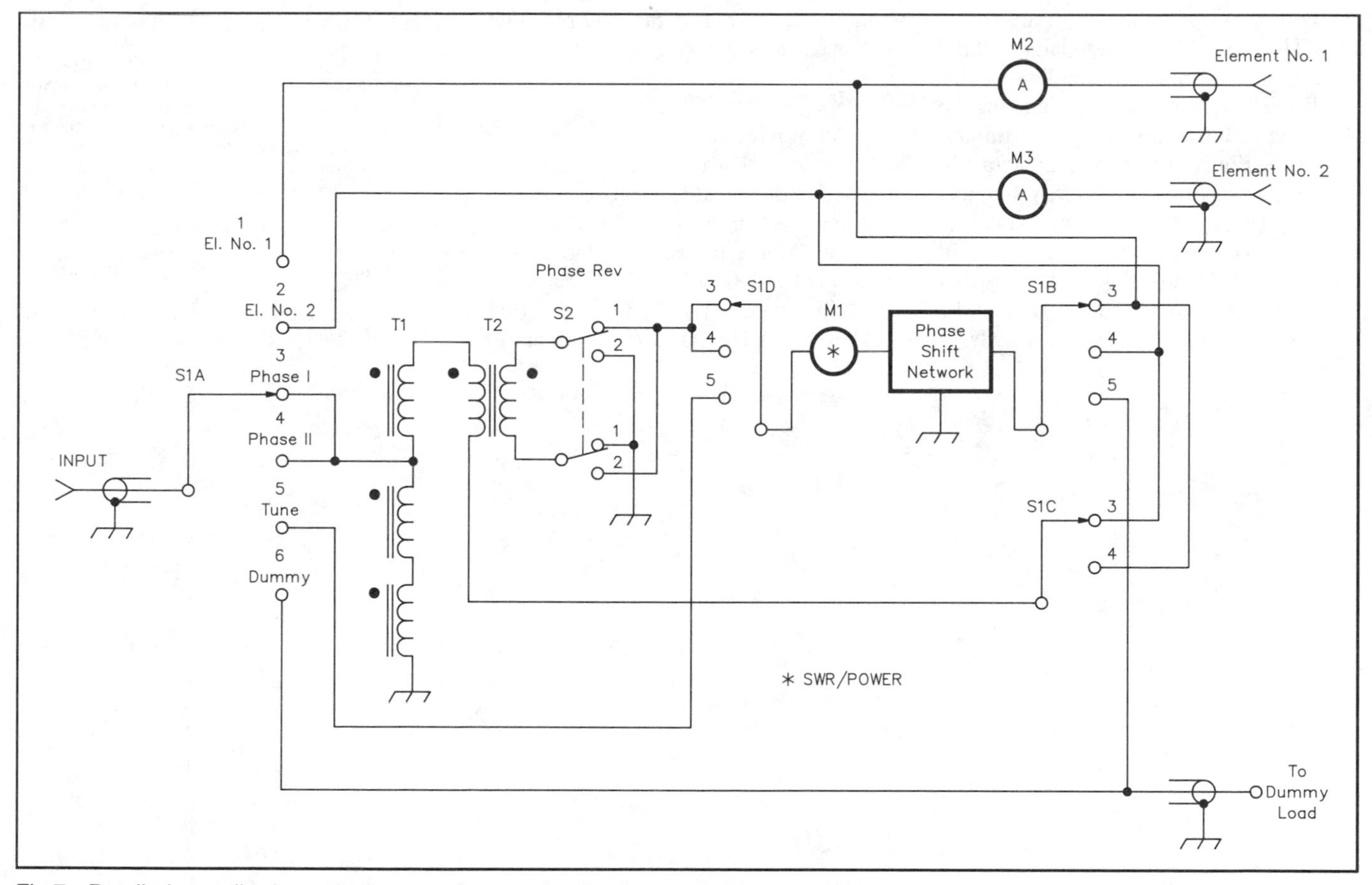

Fig 7—Detailed overall schematic diagram. See text for details on switching capabilities.

M1—SWR/power meter; see text.

M2, M3—RF ammeters; see text.

S1—Centralab PA-2013, 4 pole, 2 to 12 position rotary switch (Newark).

S2—DPDT toggle.

S3—DPST toggle.

Phase-shift network—See Figs 3 and 6.

T1, T2—See text and Fig 2.

C which gives you a low SWR. This range should be substantial for the highest L setting associated with each band, providing the ability to continuously modify the phase for a particular L setting.

With the table in hand, turn on the receiver and listen. Find a distant signal (not sky wave) and set the PHASE REV switch for whichever setting provides the weakest signal. Now switch through the various L settings, carefully adjusting capacitor C until the signal is at its absolute weakest point. Throw the PHASE REV switch to the opposite position and you should find that the signal has increased to its strongest possible level.

If you wish to work Scandinavia from the US with a beam, you might adjust the rotator to 25°, depending on where you live. This is one type of degree setting. It is important to understand that in using the phase network arrangement, you can vary the phase of one antenna by a full 360°relative to the other and, in doing so, dramatically change the directivity of the array. However, 25° of phase lag is not the same as the bearing to Scandinavia. Instead, you "steer" the array by changing the phase lag and experimentally determining the best settings to hear Scandinavia. Be sure to log the settings for future reference.

This may sound very complex, but it is not. I hunt counties on all three bands and consult a simple table for the settings that favor New England, Florida, Texas and the Northwest from my QTH here in Maryland. (Note that the settings vary from band to band.) I can switch from one band to another and be fully operational within 30 seconds.

I continuously monitor the SWR and power while utilizing the network circuitry. This fact is not emphasized in Fig 7. I repackaged a Heath HM-102 SWR/power meter into the unit (shown as M1 on the schematic) to provide this capability. Two RF ammeters (M2 and M3) were also used to continuously monitor power to each antenna. These are luxuries; you should be able to live with a single SWR and power meter between the rig and this circuit, provided you are careful when adjusting the C setting to agree with the L setting. Use of the S1 TUNE position is a great aid in quickly performing this adjustment.

Components

Note that switches S1 and S4 are relatively expensive if purchased new. You should be able to find all the switches, the variable capacitor and the 500-pF doorknob capacitors at a hamfest.

160-Meter Considerations

I do not have two 160-meter verticals and, therefore, I was not able to test the network on that band. The phase-shift network may be simply scaled by a factor of two from Table 3, the 3.5-MHz table. Note that a dual 0-2000 pF variable capacitor and an inductance of 4.72 µH are required.

The limiting factors are transformers T1 and T2. It is important that the inductive reactance associated with a winding be greater than the terminating resistance by a factor greater than five. Thus, the self-inductance must be greater than 22 µH (250 Ω at 1.8 MHz). This translates to 21 or more bifilar or trifilar windings on the Amidon R61-050-750 rods, or a total of over 60 conductor turns for T1. If this guideline is observed, the phasing arrangement should work on the top band. I would greatly appreciate any feedback on efforts to use this design on 160 meters.

Notes and References

[1]B. Alexander, "Steerable Array for the Low Bands," in this chapter.

[2]P. H. Anderson, "Impedance Matching Transformers and Ladderline," *Ham Radio*, May, 1989.

[3]L. A. Moxon, *HF Antennas for All Locations* (Potters Bar, Herts: RSGB, 1982 and 1986).

[4]The author has a number of C language (Borland Turbo C) and spreadsheets (Borland Quattro Pro) which were used to analyze the pi network and to calculate the physical characteristics of the inductors. They are available on a 5¼-inch, 360k disk in MS/DOS format. Please include $2.00 to cover postage and handling. (The ARRL in no way warrants this offer.)

[5]Amidon Inc, 3122 Alpine Ave, Santa Ana, CA 92704.

[6]Barker and Williamson, 10 Canal St, Bristol, PA 19007.

[7]Newark Electronics, Administrative Offices, 4801 North Ravenswood St, Chicago, IL 60640.

from *The ARRL Antenna Compendium, Vol 2*

The Simplest Phased Array Feed System. . .That Works

By Roy Lewallen, W7EL
5470 SW 152 Ave
Beaverton, OR 97007

Many amateurs having a phased-array antenna use the feed system shown in Fig 1, with the difference between the electrical lengths of the feed lines equaling the desired phase angle. The result is often disappointing. The reasons for poor results are twofold.

1) The phase shift through each feed line is not equal to its electrical length, and

2) The feed line changes the magnitude of the current from input to output.

This surprising combination of events occurs in nearly all amateur arrays because of the significant, and sometimes dramatic, change in element feed-point impedances by mutual coupling. The element feed-point impedances—the load impedances seen by the feed lines—affect the delay and transformation ratio of the cables. It isn't a small effect, either. Phasing errors of several tens of degrees and element-current ratios of 2:1 are not uncommon. Among the very few antennas which do work are arrays of only two elements fed completely in phase (0°) or out of phase (180°). This topic is covered in detail in *The ARRL Antenna Book*.[1]

It is possible, however, to use the system shown in Fig 1 and have the element currents come out the way we want. The trick is to use feed-line lengths which give the desired delay and transformation ratio *when looking into the actual element impedances.* More specifically, we choose the feed-line lengths to give the desired ratio of currents, with the correct relative phasing. This paper explains how to find the correct feed-line lengths. (Three BASIC programs to do the calculations is available, in either paper or electronic form, from ARRL Headquarters.[5]) Table 1 gives results for a 90°-fed, 90°-spaced, 2-element array.

Calculation of the feed-line lengths, with or without the program, requires knowledge of the element self- and mutual impedances. Most of us don't know these impedances for our arrays, so one of several approaches can be taken.

1) Measure the self- and mutual impedances using the techniques described in *The Antenna Book* (see note 1). If carefully done, this approach will lead to the best array performance.

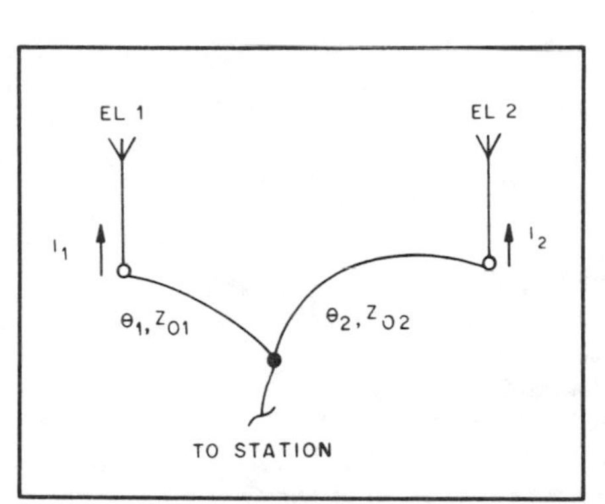

Fig 1—A typical feed system for two-element phased arrays. See text for a discussion of the feed-line lengths. Grounds and cable shields have been omitted for clarity.

2) Estimate the self- and mutual impedances. Methods and graphs are given in the *The Antenna Book*. This approach can lead to very good array performance if the array elements are straight and parallel, and with no loading elements or unusual features.

3) If the elements fit the above description, and in addition are self-resonant and close to ¼ λ high, you can use Table 1 instead of the program, if desired.

4) If you can't measure or estimate the self- and mutual impedances with reasonable accuracy, and your elements don't fit the description given in approach 3, you're likely to get poor results with this feed system. A better approach would be to use the L-network feed system described in *The Antenna Book*. It's quite simple and has the advantage of being adjustable. Adjustment methods also are given in *The Antenna Book*.

Using The Program

Program 1 was purposefully written in a very simple form of BASIC. It should run on nearly any computer without modification. If you encounter difficulty, the most likely cause is that the program was not copied exactly as printed.

The first prompts are for the self-R and X of the elements. These are the impedances which would be measured at the base of each element with the other element open-circuited at the base. The self-R includes any loss resistance. The remainder of the prompts are self-explanatory. Refer to Table 2, the sample run, for an example

Table 1

Phasing Line Lengths for a 90°-Fed, 90°-Spaced Two-Element Array

See Fig 1.

R_s*, Ohms	No. Radials per Element	Phasing Lines: Z_0, Ohms 1	2	Elect Length (Deg) Line 1	Line 2
65	4	50	50	No solution	
		75	75	No solution	
		75	50	30.36	104.96
				95.13	162.96
54	8	50	50	No solution	
		75	75	68.15	154.29
				132.60	184.95
45	16	50	50	No solution	
		75	75	58.69	153.48
				144.43	183.39
36	∞	50	50	80.56	154.53
				131.68	173.23
		75	75	51.61	155.40
				153.86	179.13

*Self-impedance, including losses.

Table 2

Sample Run of Program 1 for a 90°-Spaced, 90°-Fed Array

Calculations are for resonant elements, approximately ¼-λ high, with 8 ground radials per element.

```
RUN
SELF R, X OF LEADING ELEMENT (OHMS)
? 54,0
SELF R, X OF LAGGING ELEMENT (OHMS)
? 54,0
MUTUAL R, X (OHMS)
? 20,-15
EL.2:EL.1 CURRENT MAGNITUDE, PHASE (DEGREES)
—PHASE MUST BE ZERO OR NEGATIVE
1,-90
FEEDLINE 1, 2 IMPEDANCES (OHMS)
? 50,50
NO SOLUTION FOR THE SPECIFIED PARAMETERS.
  WOULD YOU LIKE TO TRY DIFFERENT
  FEEDLINE Z0'S (Y,N)? Y
FEEDLINE 1, 2 IMPEDANCES (OHMS)
? 75,75
                  Z0= 75 OHMS          Z0= 75 OHMS
                  TO LEAD. EL.         TO LAG. EL.
                  ELECT. L. (DEG.)     ELECT. L. (DEG.)
FIRST SOLN.       132.6038             184.9522
SECOND SOLN.      68.1518              154.2918
Ok
```

of program operation.

Sometimes you might get the result, NO SOLUTION FOR THE SPECIFIED PARAMETERS. This doesn't mean there's a solution which the program couldn't find; it means that there really is no solution for the specified conditions. If this happens, try different feed-line impedances. I've found a combination of common feed-line impedances which will work with nearly every array I've wanted to feed, but there are some which can't be fed using this method.

Whenever there is a solution, there's also a second one. Both are computed by the program. It may be necessary to use the longer set of feed-line lengths in order to make the feed lines physically reach the elements. You can also add ½ λ of cable to *both* feed lines and maintain correct operation. For example, the array in the sample program run of Table 2 can be fed with two 75-ohm lines of the following lengths (given in electrical degrees).

68.15° and 154.29°
132.60° and 184.95°
248.15° and 334.29°
312.60° and 364.95° (or 4.95°)

The first two sets are the lengths given by the program. A half wavelength is added to both lines to make sets 3 and 4. Note that a *full* wavelength can be subtracted from the second line length in the last set.

Occasionally it's necessary to make the feed-line impedances different from each other. If you want to be able to switch the pattern direction but have unequal feed-line impedances, add ½ λ of line from each element to the phasing feed line. If both ½-λ lines have the same impedance, directional switching will be possible while maintaining correct phasing.

Using the Table

Table 1 gives the feed-line lengths necessary to correctly feed a 90°-fed, 90°-spaced, 2-element array. The table is based on the following assumptions:

1) The elements are identical and parallel.

2) The ground systems of the elements have equal loss.

3) The elements are resonant when not coupled to other elements. A height of $237/f_{MHz}$ will be close to resonance for most vertical elements.

4) The elements are not loaded and do not have matching networks at their bases. Traps generally act like loading elements on the lower bands.

5) The elements are fairly "thin." HF antennas made from wire, tubing, or common tower sections fit this category.

6) Your ground isn't unusually dry or swampy. If it is, you may have more or less element self-resistance than shown for the number of radials. The resistance versus number of radials is based on measurements by Sevick.[2]

Since so many factors can affect ground losses and element self- and mutual impedances, the tables probably won't give exactly the best feed-line lengths for your array. But if the above assumptions apply, it's very likely that your array will work better using the recommended feed-line lengths. If the assumptions don't describe your array, the table values won't be valid.

Two Four-Element Arrays

The Antenna Book (note 1) describes a feed system for two types of four-element arrays based on a combination of the "current forcing" method and an L network. Information on these arrays, the current forcing method, and practical advice on how to measure the various line sections can be found in Chapter 8 of *The Antenna Book*. The L network can be replaced by two feed lines, resulting in the feed systems shown in Figs 2 and 3. The principle is the same as for the two-element array, although the mathematics are a bit different due to the presence of the λ/4 or 3 λ/4 lines and the difficulty of including the mutual impedances between all elements. The mathematics are described in the next section.

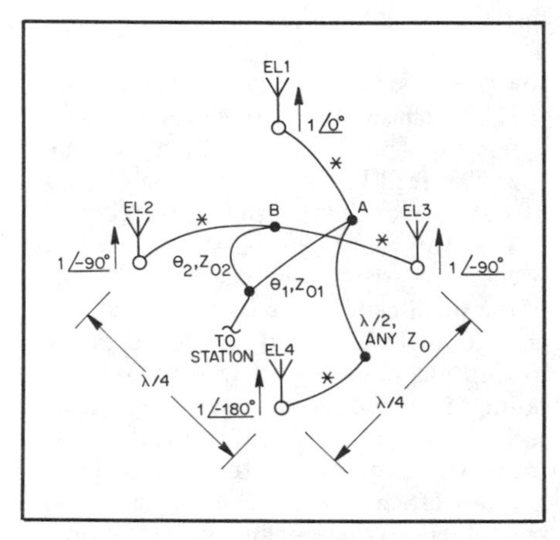

Fig 2—Feed system for the four-square array. Grounds and cable shields have been omitted for clarity. The lines marked "*" are all the same length, have the same Z_0, and are electrically either λ/4 or 3λ/4 long. The other lines are discussed in the text.

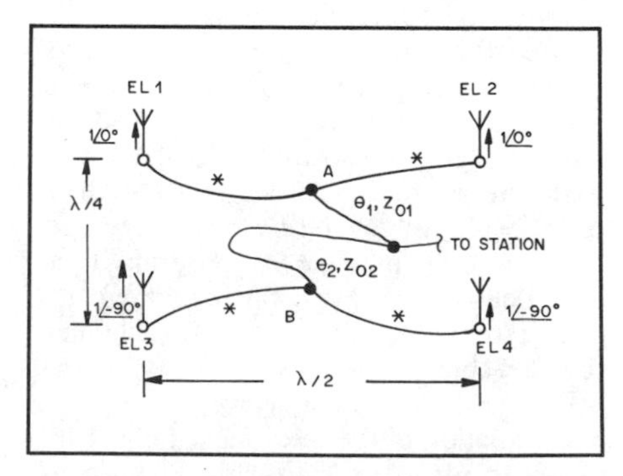

Fig 3—Feed system for a four-element rectangular array. Grounds and cable shields have been omitted for clarity. The lines marked "*" are all electrically 3λ/4 long and have the same Z_0. The other lines are discussed in the text.

Tables 3 and 4 give feed-line lengths for these two arrays. The same restrictions apply to the four-element tables as to the two-element table. They were calculated using modified versions of the BASIC program. These programs, which apply only to the four-square and rectangular arrays, are listed as Programs 2 and 3.

The Mathematics

For an array to work properly, the element currents need to have the correct relationship. So let's first look at the general problem of feeding two loads with a specific ratio of currents (Fig 4). The desired current ratio, I_2/I_1, is a complex number with two parts: magnitude M_{12} and angle ϕ_{12}. Both parts must be correct for the array to work as planned. Assuming for the moment that we know what the load impedance will be, we can write the following equation for feed-line no. 1.

$$V_{in} = I_1 Z_1 \cos\theta_1 + jI_1 Z_{01} \sin\theta_1 \quad \text{(Eq 1)}$$

where

V_{in} = voltage at the input end of the line

I_1 = current at the output end of the line

Z_1 = load impedance at the output end of the line

θ_1 = electrical length of the line in degrees or radians

Z_{01} = characteristic impedance of the line

V_{in}, I_1, and Z_1 are complex

This is the general equation which relates the output current to the input voltage for a lossless transmission line.[3] A similar equation can be written for the second feed line. Since the feed lines are connected together at their input ends, the input voltages are equal, and we can write

$$V_{in} = I_1 Z_1 \cos\theta_1 + jI_1 Z_{01} \sin\theta_1$$

$$V_{in} = I_2 Z_2 \cos\theta_2 + jI_2 Z_{02} \sin\theta_2$$

Rearranging to solve for the current ratio gives

$$\frac{I_2}{I_1} = \frac{(Z_1 \cos\theta_1 + jZ_{01} \sin\theta_1)}{(Z_2 \cos\theta_2 + jZ_{02} \sin\theta_2)} \quad \text{(Eq 2)}$$

This equation can be used to illustrate the problems of feeding unequal load impedances (present in the elements of most arrays). For example, if values that might be found in a 90°-spaced, 90°-fed array are

$Z_1 = 35 - j20\ \Omega$
$Z_2 = 65 + j20\ \Omega$
$Z_{01} = Z_{02} = 50\ \Omega$
$\theta_1 = 90°$, and $\theta_2 = 180°$

then I_2/I_1 would be 0.735 at an angle of −107°, not 1 at an angle of −90° as planned. In a real array, because of mutual coupling, the element feed-point impedances are modified by the currents

Table 3

Phasing Line Lengths for a Four Square Array

See Fig 2.

R_s,* Ohms	No. Radials per Element	λ/4 Line Z_0, Ohms	Phasing Lines: Z_0, Ohms A	Z_0, Ohms B	Elect Length (Deg) Line A	Line B
65	4	50	50	50	20.66	166.50
					147.94	204.91
			75	75	13.70	170.60
					158.00	197.70
		75	50	50	No solution	
			75	75	32.03	162.26
					133.53	212.18
54	8	50	50	50	25.82	166.32
					138.25	209.61
			75	75	16.80	170.01
					151.13	202.06
		75	50	50	No solution	
			75	75	45.11	167.22
					115.95	211.72
45	16	50	50	50	34.57	168.71
					123.73	212.98
			75	75	21.15	170.17
					141.45	207.23
		75	50	50	22.36	121.53
					134.77	261.34
			75	75	No solution	
36	∞	50	50	50	No solution	
			75	75	31.37	173.66
					121.79	213.18
		75	50	50	33.55	122.94
					120.01	263.50
			75	75	No solution	

*Self-impedance, including losses.

Table 4

Phasing Line Lengths for a Four Element Rectangular Array

See Fig 3.

R_s,* Ohms	No. Radials per Element	3λ/4 Line Z_0, Ohms	Phasing Lines: Z_0, Ohms A	Z_0, Ohms B	Elect Length (Deg) Line A	Line B
65	4	50	50	50	37.37	155.34
					132.87	179.10
			75	75	24.95	162.22
					150.41	181.47
		75	50	50	No solution	
			75	75	66.82	153.30
					87.34	161.46
54	8	50	50	50	59.02	151.97
					100.15	163.93
			75	75	35.79	157.33
					135.17	175.44
		75	50	50	No solution	
			75	75	No solution	
			75	50	29.44	112.55
					67.01	129.30
45	16	50	50	50	No solution	
			75	75	62.38	152.17
					98.08	159.50
		75	50	50	No solution	
			75	75	No solution	
			75	50	12.61	99.74
					64.34	114.66
36	∞	50	50	50	No solution	
			75	75	No solution	
		75	50	50	52.04	99.08
					176.88	274.80
			75	75	No solution	

*Self-impedance, including losses.

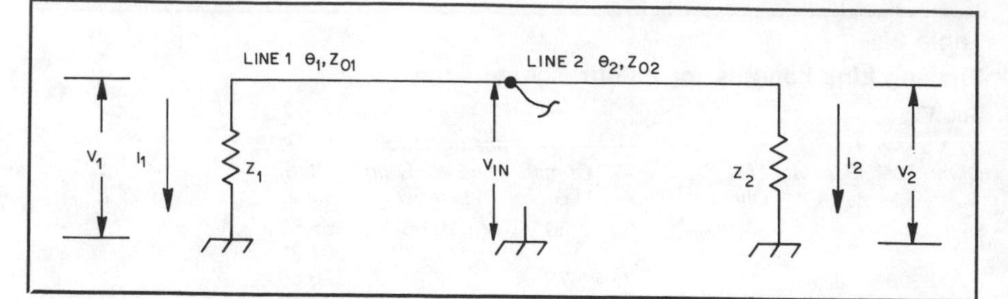

Fig 4—Feeding two load impedances with specific currents. This example assumes that Z1 and Z2 are not affected by mutual coupling.

flowing in the elements. But the element currents are a function of the element feed-point impedances, so Eq 2 can't be used directly to calculate currents in array elements. To write an equation which will do that, we need to modify Fig 4 to account for the effect of mutual coupling (Fig 5). From the diagram,

$$V_1 = I_1 Z_{11} + I_2 Z_{12} \quad \text{(Eq 3)}$$

$$V_2 = I_2 Z_{22} + I_1 Z_{12} \quad \text{(Eq 4)}$$

where

V_n = voltage at the feed point of element n

I_n = current at the feed point of element n

Z_{nn} = self-impedance of element n (the feed-point impedance when the element is totally isolated from all other elements)

Z_{12} = mutual impedance between the elements

All variables are complex

A slightly different form of Eq 1 is

$$V_{in} = V_1 \cos\theta_1 + jI_1 Z_{01} \sin\theta_1 \quad \text{(Eq 5)}$$

and for the second feed line

$$V_{in} = V_2 \cos\theta_2 + jI_2 Z_{02} \sin\theta_2 \quad \text{(Eq 6)}$$

V_1 and V_2 from Eqs 3 and 4 are substituted into Eqs 5 and 6. The right sides of Eqs 5 and 6 are set equal to each other, since V_{in} is the same for both feed lines. Finally, the resulting equation is rearranged to solve for I_2/I_1.

$$\frac{I_2}{I_1} = \frac{Z_{11}\cos\theta_1 - Z_{12}\cos\theta_2 + jZ_{01}\sin\theta_1}{Z_{22}\cos\theta_2 - Z_{12}\cos\theta_1 + jZ_{02}\sin\theta_2} \quad \text{(Eq 7)}$$

This is the same as Eq 2 except that an additional term containing mutual impedance Z_{12} appears in both the numerator and denominator. Given the element self- and mutual impedances and the lengths and impedances of the feed lines, Eq 7 can be used to find the resulting ratio of currents in the elements. The problem we're trying to solve, though, is the other way around: how to find the feed-line lengths, given the current ratio and other factors. Christman described an iterative method of using Eq 7 to solve the problem by beginning with an initial estimate of feed-line lengths, finding the resulting current ratio, correcting the estimate, and repeating until the answer converges on the correct answer.[4] This method gives accurate answers, and I used it for some time. The disadvantage of the approach is that convergence can be slow, and the iterations can actually diverge for some arrays unless the program includes "damping."

Fortunately, an iterative approach isn't necessary, since Eq 7 can be solved directly for feed-line lengths. The method is straightforward, although tedious, and was done using several variable transformations to keep the equations manageable. The details won't be described here. The BASIC programs (mentioned earlier and in Note 5) use the direct solution method, and the validity of the results can be confirmed by substitution into Eq 7.

The feed system can be adapted to certain larger arrays by combining it with the current forcing method described in *The Antenna Book* (see Figs 2 and 3). The basic requirement is to make the *voltages* at points A and B have the proper ratio and phase angle. If this is accomplished, the elements will have correct *currents* because of the properties of the ¼-λ lines. *The Antenna Book* shows the use of an L network to obtain the voltage phase shift; the same thing can be accomplished by using two feed lines of the correct length.

To see how we can use the program to solve the problem, we'll rewrite Eq 2 to apply to the currents and impedances at points A and B:

$$\frac{I_B}{I_A} = \frac{Z_A\cos\theta_1 + jZ_{01}\sin\theta_1}{Z_B\cos\theta_2 + jZ_{02}\sin\theta_2} \quad \text{(Eq 8)}$$

Because $V_A = I_A Z_A$ and $V_B = I_B Z_B$, then

$$\frac{V_B}{V_A} = \frac{Z_B}{Z_A}\frac{I_B}{I_A} = \frac{Z_B}{Z_A}\frac{Z_A\cos\theta_1 + jZ_{01}\sin\theta_1}{Z_B\cos\theta_2 + jZ_{02}\sin\theta_2} \quad \text{(Eq 9)}$$

Note the similarity to Eq 7, which is the equation the program solves for θ_1 and θ_2. We can use the program to solve Eq 9 if we

1) Enter Z_A when it prompts for the self-Z of element 1,

2) Enter Z_B when it prompts for the self-Z of element 2,

3) Enter 0,0 when it prompts for the mutual R, X, and

4) Enter $(V_B/V_A)\,(Z_A/Z_B)$ when it prompts for the desired current ratio. For the two four-element arrays,

$$V_B/V_A = 0 - j1 = 1\,\angle{-90°}.$$

The following steps are required to calculate Z_A/Z_B.

1) Measure or estimate the self- and mutual impedances of the elements.

2) Using the self- and mutual impedances and the current ratios, calculate the actual element feed-point impedances. The method is described in *The Antenna Book.*

3) Calculate the impedances looking into the λ/4 or 3λ/4 lines.

4) Where two of the lines are connected,

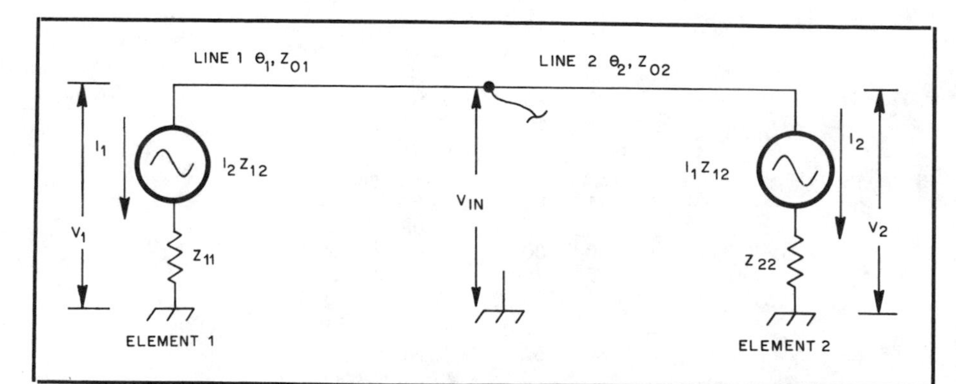

Fig 5—Feeding two antenna elements with specific currents. The voltage sources are added to account for mutual coupling.

calculate the parallel impedance. These will be Z_A and Z_B.

5) Calculate the ratio Z_A/Z_B.

Program 1 has been modified to do these calculations for you. The modified BASIC programs are listed as Programs 2 and 3. [All three programs are available on 5.25-inch diskette for the IBM PC and compatibles; see information in Note 5.—*Ed.*] You must know the self-impedance of an element (all are assumed to be identical) and the mutual impedances between all elements in order to use the modified programs.

Closing Comments

I first solved Eq 7 for the feed-line lengths several years ago. However, I didn't try to publish the results because of the large amount of explanation which would have to go with it—why the common feed method frequently is disappointing, and explaining the current forcing and L-network feed systems, the role of mutual coupling in phased arrays, etc. I want to thank Jerry Hall, K1TD, for providing the opportunity to explain them in a forum which is readily available to amateurs —Chapter 8 of *The ARRL Antenna Book*.

Your array *will* work better if properly fed. This feed system isn't any more complicated than the one you've probably been using, but it's likely to give you much better results. Try it!

Notes

[1]G. L. Hall, Ed, *The ARRL Antenna Book*, 15th ed. (Newington: ARRL, 1988), Chap 8.

[2]J. Sevick, "The Ground-Image Vertical Antenna," *QST*, Jul 1971, pp 16-19, 22. Also "The W2FMI Ground-Mounted Short Vertical," *QST*," Mar 1973, pp 13-18, 41. (Summary information from these articles is presented graphically in Fig 23, p 8-23 of *The Antenna Book*—see note 1).

[3]*Reference Data for Radio Engineers*, 5th ed. (Howard W. Sams & Co, 1968).

[4]A. Christman, "Feeding Phased Arrays: An Alternative Method," *Ham Radio*, May 1985, p 58 and Jul 1985, p 74.

[5]Programs 1, 2 and 3 are on the optional companion disk for *The ARRL Antenna Compendium, Volume 2*, which may be purchased for $10 from ARRL Headquarters. Printed listings of the programs are available from the ARRL Technical Department Secretary, 225 Main St, Newington, CT, 06111-1494. Send an SASE and ask for "Vertical Anthology Lewallen Programs."

from September 1981 *QST* (Hints and Kinks)

SWITCHING 40-METER PHASED VERTICAL ANTENNAS

☐ During one of our QSOs, Dick Evans, VE6XW, of Millet, Alberta, explained the antenna-switching arrangement he has for his 40-meter phased vertical antennas. The method he devised stems from his vocation and expertise as an electrician. He pointed out that his research led him to believe that at least some parts of his design have not been presented by technical writers in Amateur Radio publications.

His system (Fig. 3) provides six end-fire unidirectional selections for 40 meters. It consists of three identical base-loaded vertical radiators. Dick suggests the use of 5/8-wavelength antennas, but points out that 30-foot (9-meter) elements with loading coils are satisfactory. He used a design frequency of 7.100 MHz.

Feed lines W1, W2 and W3, connecting the switch (S1) with the antennas, have identical lengths of 52-ohm coaxial cable. Cables W4, W5 and W6 are each 22 feet, 10 inches (7 m) long. They are neatly coiled indoors. Direction changes are accomplished by a six-position isolated double-pole switch (S1) located at the operating desk. For QRP transmitters, a Radio Shack no. 275-1386 switch is satisfactory; for higher power, a Millen transmitting type of switch (or equivalent) is suggested.

A minimal radial system would consist of four 1/4-wavelengths of wire for each vertical, but 10 to 20 radials per vertical would be better. With this system, excellent DX results can be expected.

Dick has agreed that he would like to share this information with other amateurs. I wish to thank him for letting me present this to *QST*. — *Chuck Coleman, K6ZUR, Santa Rosa, California*

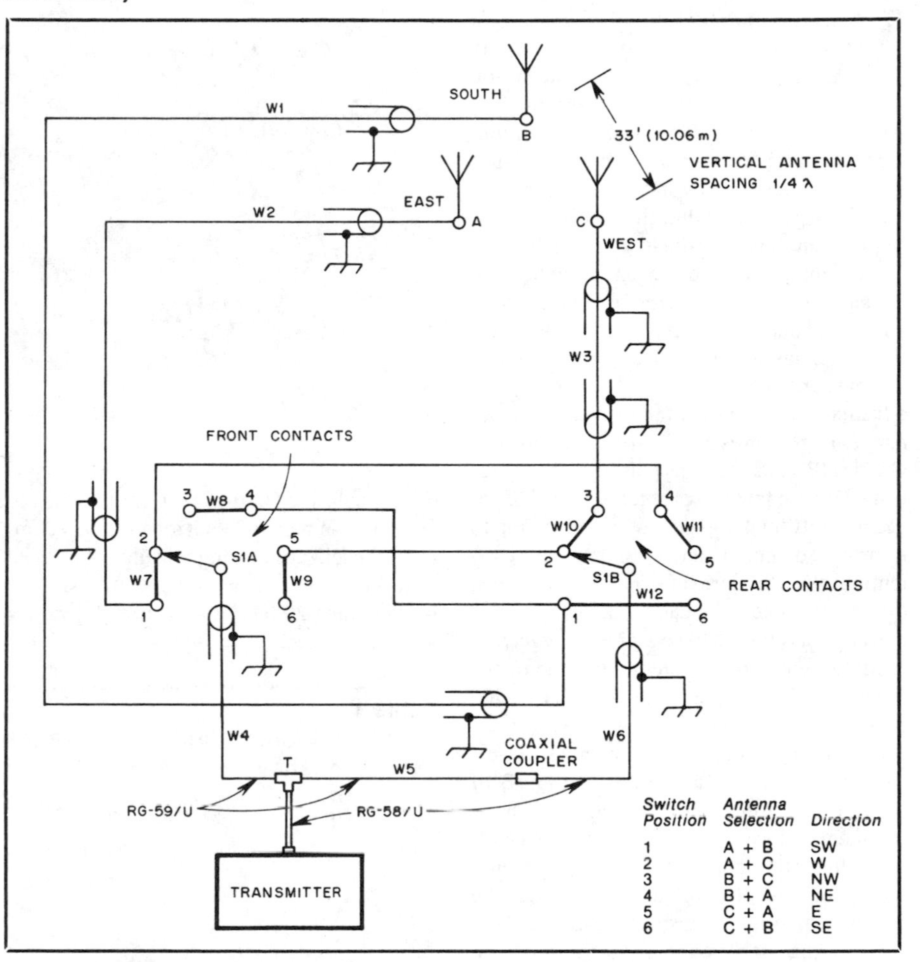

Fig. 3 — This antenna-switching arrangement used by Dick Evans, VE6XW, provides six end-fire unidirectional antenna selections for his phased vertical antenna array. Although it is designed for use on 40 meters, it can be adapted to other bands and broadside arrays. Jumper wires and connecting cables in the drawing are identified by the letter W.

The Tuned Guy Wire–Gain for (Almost) Free

By John Stanley, K4ERO
8495 Hwy 157
Rising Fawn, GA 30738

K4ERO describes a practical low-frequency vertical antenna with gain and good pattern. You may be able to modify an antenna you already have in your backyard.

With the bottom of the sunspot cycle fast approaching, many hams are turning their attention to the "low bands"—40, 80 and 160 meters. As low band DXers know, the low frequencies do not go dead when we have low sunspots—they actually work somewhat better. Now is the time to be thinking about improving your antennas for the lower frequencies.

I'm going to describe an antenna that doesn't use a new technique, but a much underused one: the *tuned guy wire*. Most vertical antennas need guy wires for mechanical reasons, and these guys can be used to your advantage. In the past, attempts to avoid resonances in guy wires may have been overdone. On the HF bands it seems that solid guy wires usually cause few real problems.[1] The idea of deliberately making guy wires resonant to improve antenna patterns deserves more attention. With modern antenna analysis tools, this has become quite easy to do.[2]

Let us assume that you presently have a pretty decent vertical antenna for one of the lower bands, with a decent buried-radial ground system. Even if your vertical is not so decent, keep reading. You can get a front-to-back ratio of 20 dB and more than 3 dB of forward gain from your vertical, with an investment of a weekend's work and a few dollars! Think what that front-to-back ratio would do to your ability to work the weak ones in the presence of strong stateside QRM.

Changing one of your present non-resonant guys to a tuned reflector can accomplish this magic. The following is for 3.8 MHz, but you can scale all dimensions for either 160 or 40 meters. See **Table 1**. Measure out from the base of your vertical about 50 feet. Put down some kind of anchor in the ground. Prepare a wire 60.2 feet long and hang it between the ground anchor and the top of your tower, insulating the wire at both ends. It won't reach, you say. Good. That means that your tower is tall enough, say 50 to 70 feet, for this plan to work. (I assumed a 63-foot tower for my computer model.) Add enough insulated material to reach the top of the tower. Phillystran is ideal and the 20 feet or so you will need for this project should cost under $15.[3] Use any handy plastic rope if you wish, but plan to replace it every 2 to 3 years.

Table 1
Length of Slanted Reflector Wire for Low-Frequency Amateur Bands

MHz Freq	*Reflector Length*
1.8	127 ft 1 inch
1.9	120 ft 4 inches
2.0	114 ft 5 inches
3.6	63 ft 6 inches
3.8	60 ft 2 inches
7.1	32 ft 2 inches
7.3	31 ft 4 inches

Add insulated rope or Phillystran to wire to reach to a point near the top of your vertical antenna. Use a mechanically sound insulator between anchor point and bottom of the tuned wire. Connect the bottom end via a flexible lead through the loading inductor and then to the ground radials.

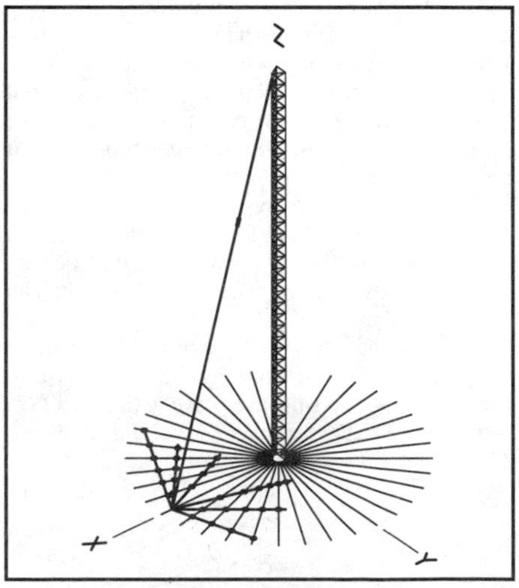

Fig 1—Bird's eye view of K4ERO tuned guy wire technique. Note that ground radials used by the vertical driven element are "shared" by slant-wire reflector element.

Table 2
Coil Specifications for Low-Frequency Amateur Bands

Freq MHz	*Turns*	*Wire size AWG*	*Dia inches*	*Length inches*
1.9	12	14	3	3
3.6	8	14	3	2
7.1	5	14	3	1.5

All dimensions in inches. Other combinations of turns and sizes are also usable, but the coil should not be physically too small if power capability and Q are to be maintained. X_L of coil to be maximum of 100 Ω; only a portion will be used in final antenna, based on actual results of tuning.

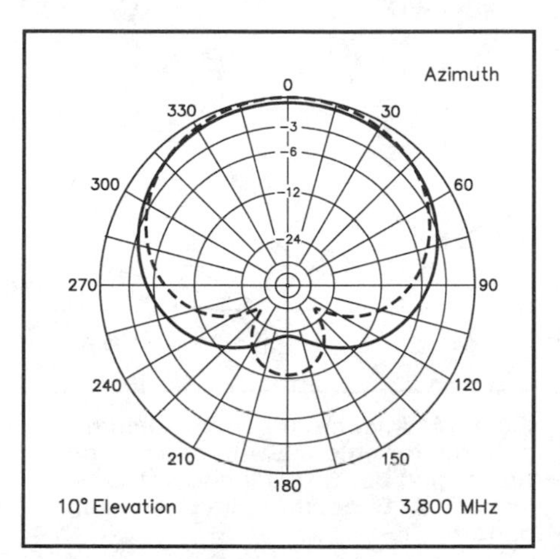

Fig 2—Azimuth patterns for 80-meter K4ERO array, with null adjusted at 180° azimuth by reflector tuning (solid line). Null can be placed in a different direction by retuning reflector, as shown by dashed line.

Cut slits in the ground from the anchor point to uncover your radials in 6 different directions. See **Fig 1**. At each point where you cross an existing radial, make a solder connection to a #12 bare wire you have laid in the slit. Tie all the new radials together at the anchor point. Between the radials and the bottom of the slanted wire, connect a coil. This should be adjustable up about +100 Ω of reactance. See **Table 2** for design of the coil. Apply a low-power signal to the vertical and measure the field strength in the direction of the sloping wire. Adjust the coil for a minimum signal. A null of between 10 and 20 dB should be attainable, depending mainly on your radial system. An even better method is to use a small signal source placed in the direction you wish to null and tune the reflector for minimum received signal. You may wish to tune for a null at an angle other than directly off the back. For example, you may have an interference source (line noise, etc) at an angle between 180° (off the back) and 90°. This can be dropped into a pattern null by tuning for the dashed-line pattern shown in **Fig 2**. I have done this on an actual antenna with results very close to those calculated. (*NEC* modeling really does work.)

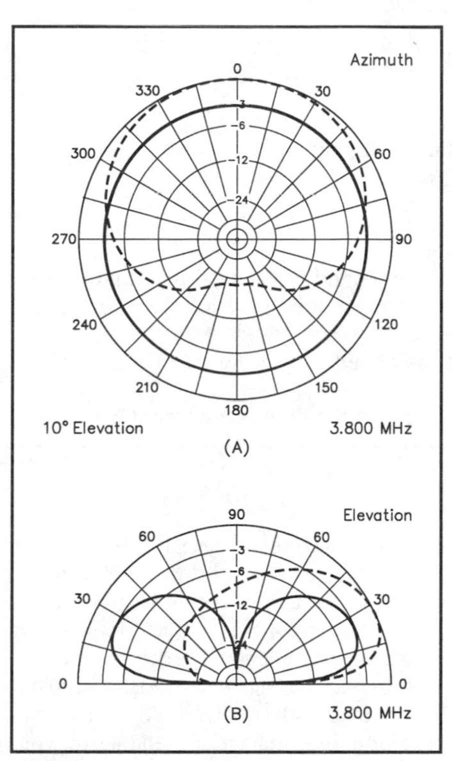

Fig 3—At A, comparison of azimuth patterns for antenna with and without tuned-guy reflector. The radial system is assumed to be excellent in this computation. At B, elevation patterns are shown.

When the desired null is obtained, go to the matching network for the vertical. You should find that the SWR has changed somewhat from what it was before you installed the tuned reflector, perhaps by as much as two to one. This is good, indicating that the sloping wire is indeed affecting the vertical, as it should. You should find that the input impedance has gone more inductive because the tuned radial has made the vertical look longer. This means that you will have to compensate by removing some inductance, adding some series capacity, or in some other way retuning your matching network at the base of the vertical. Of course, if you have a tuner in the rig or shack, you may just want to reset it a bit, as a 2:1 SWR will introduce negligible loss in your system, unless you already have far too much loss in your feed line.

At this point, you may want to check the forward gain. It should be 1.5 to 3.5 dB, again depending on your radial system. See **Fig 3** for a comparison of antenna with and without slanted reflector. Do not check the gain without retuning and readjusting the power of your transmitter, as its output may have fallen off while you were tuning for F/B ratio.

If you have an easy way to disconnect the tuned guy wire from its loading coil, you can quickly go back to the omnidirectional pattern. You may also wish to install a few other tuned guys for other directions and select the one you wish by connecting it to its loading coil, while leaving the others open. The solid 60.2-foot wire will not affect the pattern, provided it is left unconnected at the bottom end.

Very importantly, let me mention that unless you have a lot of experience in tower work and know how to use temporary guys when modifying existing ones and know how to tell when a guy is properly tensioned, or if you have an experienced friend who will work with you to do this project, I recommend that you use separate wires for the reflectors, and DO NOT use the steel guy wires that HOLD UP your tower. If your tower is not self supporting, do not disconnect any of the guys, even temporarily, lest the tower fall. If your tower is normally unguyed, the addition of 3 or 4 tuned "guys" will give it a bit of extra strength for high winds, etc.

Do not compromise the structural soundness by any attempts to get gain. It isn't worth it. Use strong wire, such as the "silky" #12 wire sold by the "Wireman," to help make for a safe installation.[4] Soft copper wires will stretch, and you will have to retune your reflectors from time to time.

Table 3 will give you an idea of what to expect as you tune your guys. Reflector tuning is highly recommended, since it is easy to tune for F/B, which occurs close to the tuning point needed for good gain. Tuning as a director is not recommended, since it tends to lower the radiation resistance of the driven tower. Unless you have an exceptionally good ground system, this can do more harm than good. The ground loss appears as a resistance in series with the radiation resistance, and this will cause the efficiency to drop quickly. If you have only a modest ground system, do not despair. Reflector tuning can actually raise the radiation resistance somewhat, and the effect on pattern is not nullified, just reduced.

Fig 4 illustrates the azimuth and elevation patterns to be expected for a ground system having a series loss equal to 10 Ω, represent-

Table 3
Effect of Coil Tuning on Performance

X_L Ω	R_r Ω	X_s Ω	*Gain* *dBi*	*F/B* *dB*	
0	26.2	17.7	1.78	3.0	
5	25.2	18.5	1.91	3.2	
10	24.1	19.5	2.04	3.3	
15	22.8	20.8	2.19	3.5	Director tuning
20	21.8	22.4	2.35	3.5	of guy wire
25	20.0	24.7	2.48	3.4	
30	18.5	27.7	2.56	3.1	
35	17.2	31.8	2.45	2.1	
40	16.7	37.4	1.89	0.4	Bidirectional (Fig 3A)
45	18.1	44.7	2.63	2.3	
50	23.0	52.8	3.39	6.3	
55	32.6	59.2	3.55	12.4	
60	45.3	60.0	3.32	25.2	
65	56.3	54.0	2.95	17.7	Reflector tuning
70	62.3	46.3	2.59	12.0	of guy wire
75	64.3	38.3	2.27	9.2	
80	63.9	32.1	2.00	7.4	

Above values are for a quarter-wave vertical with a good radial system. Other sizes and radial systems will give different results, but in general the results will be similar.
R_r = radiation resistance at base of fed vertical.
X_s = series inductive reactance as seen at base of vertical.
F/B = front to back ratio at 0° and 180°.
(Note that main lobe changes direction as X_L used in tuning passes through 40 Ω.)

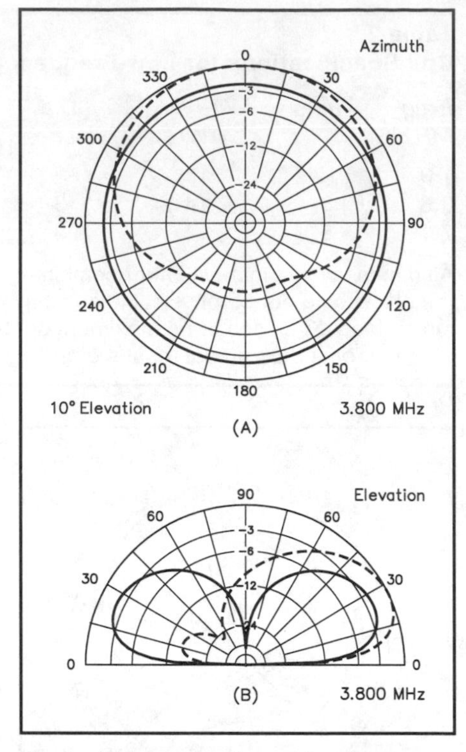

Fig 4—At A, comparison of azimuth patterns for antenna with and without tuned-guy, but with a limited radial system. At B, comparison of elevation patterns.

ing a radial system of 24 radials about 1/8 wave long. Compare these results with the patterns to be expected with a very good radial system as seen in Fig 3. Some may question how using only 6 radials for the reflector can avoid excessive loss. It should be observed that the function of radials is to shield the E-fields produced by the wires from causing currents to flow in lossy ground. This shielding is made possible by a wire spacing that is small in terms of a wavelength. From any given point on the ground's surface, the induced currents can "find" a nearby wire to flow in, rather than flowing in the resistive earth. If the wire it finds is going at right angles to its desired direction of flow, it will follow that wire until it finds a wire going in the right direction, that is, toward the base of the wire or tower from which the original E-field came. Thus, the six widely spread wires are "borrowing" the original extensive radial system to provide return paths.

Radials on the "back" side of the reflector are not very important for two reasons. First, the slope of the reflector makes the E-fields much stronger on the front side of the reflector, and second, the lack of radials in the back direction may actually contribute to the F/B ratio. This is another reason why the reflector tuning is preferred to the director tuning. In the director case, it would seem to be desirable to add additional radials in the direction of the beam; that is, beyond the tuned guy and away from the driven tower.

When using a sloping reflector, you will find that the usual null of a standard vertical at 90° elevation is lost. This could be an advantage or a disadvantage, depending on your needs. If you have a lot of potential QRM from within a few hundred miles of your QTH, you may find that the signals from short skip, which your present vertical rejects, are now a problem. On the other hand, if you use the lower bands for local net operation or rag-chewing as well as DX, you may be able to take down your low band dipole! Only very close-in signals, say a hundred miles from your QTH, will be increased by this change in the response at higher angles. For stations more than a few hundred miles away, the reduction in QRM off the back should be very noticeable.

The changes in patterns and SWR with frequency are shown in **Fig 5**. Since most low-band DXers use a relatively small portion of the band, retuning may not be necessary, although changing from 80-meter CW to 75-meter SSB will require you to readjust the base coils on your reflectors for best results.

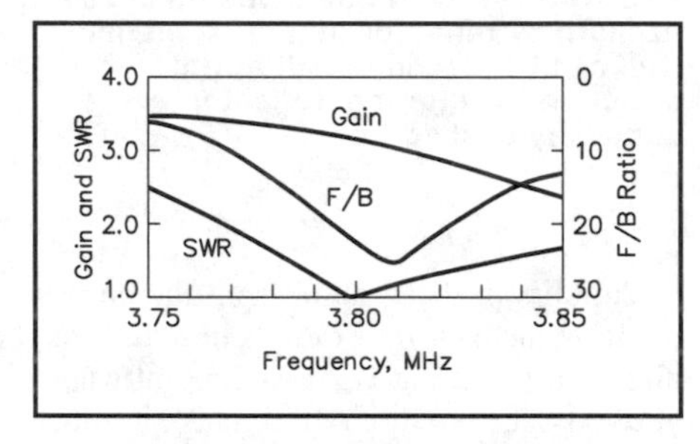

Fig 5—Variation of SWR, gain and F/B vs frequency across 75-meter DX band, excellent radial system assumed.

Notes

[1] Lew Gordon, K4VX, "The Effect of Continuous, Conductive Guy Wires on Antenna Performance," *QST*, Aug 1993, p 22.

[2] The graphics and tables included in this article were computed using *MN*, version 4.5, by K6STI. Other implementations of *NEC* and *MININEC* show similar results.

[3] See *The ARRL Antenna Book*, 15th edition, p 3-13 or *QST*, Dec 1976, p 13.

[4] The Wireman can be reached at 800-433-9473. Phillystran is sold by Texas Towers, 800-272-3467. Both sources are frequent *QST* advertisers.

Chapter 5

Reduced Size

from March 1990 *QST*

Evolution of the Short Top-Loaded Vertical

If a quarter-wave vertical for 160 meters is beyond your means, how about trying something shorter?

By Charles J. Michaels, W7XC
13431 N 24th Ave
Phoenix, AZ 85029

On 160 meters, "the gentleman's band," many hams operate short base- or top-loaded verticals. They use an antenna length (height) of about 30 to 60 feet with modest radial systems. Their available space, height restrictions or finances simply do not permit installing the traditional ¼-λ tower and a system of 120 ¼-λ radial wires that approaches an efficiency of 100%. The name of the game is, "Make the most of what you have."

For purposes of discussion, I will use a 40-foot mast of 1.6-inch average diameter and a frequency of 1.9 MHz as the basis for various short vertical antennas. Differences between the field pattern of such a short vertical and that of a ¼-λ vertical over the customary "perfect earth" are almost indistinguishable unless the patterns are superimposed. See Fig 1A. The ¼-λ radiator is very slightly better at the lower angles and the 40-foot radiator is very slightly better at the higher angles.

Over real earth, with enough input power to each antenna to produce equal *radiated* power, the patterns will again be essentially the same, and look like those of Fig 1B. Aside from questions of efficiency, the antenna patterns should be similar for the same location.

Over real earth with a modest radial system of 10 to 20 radials, each of perhaps 35 or so feet in length, antennas in various locations will not all see exactly the same ground-loss resistance. For purposes of discussion, let's assume it to be 15 ohms—recognizing that the typical short antenna is seldom found on the "average earth," that is, in a meadow accommodating an extensive radial system. This figure is derived largely by experience with suburban backyard antenna systems. This *ground-loss resistance* (R_g) will appear at the feed point of each antenna in series with the *radiation resistance* (R_r) and any other loss resistance.

For those interested in the mathematics or in designing for a somewhat different antenna height, etc, the equations used in this discussion are given in the Appendix. There are many formulas and curves for radiation resistance, R_r, versus vertical height of the antennas discussed here. Accurate formulas are tedious of solution (see Appendix Eq 1) and accurate simple formulas are restricted to narrow height ranges. Appendix Eq 2, however, applies to simple vertical monopoles with acceptable accuracy for our purposes up to heights of 90°. R_r is virtually independent of radiator diameter within any practical diameter range of these antennas.

The various antennas to be discussed are shown in Fig 2. The sinusoidally distributed currents are shown to scale. The value of the base current for 100 W input to the antenna is listed as I_b. At 1 kW, it would be 3.16 times that value.

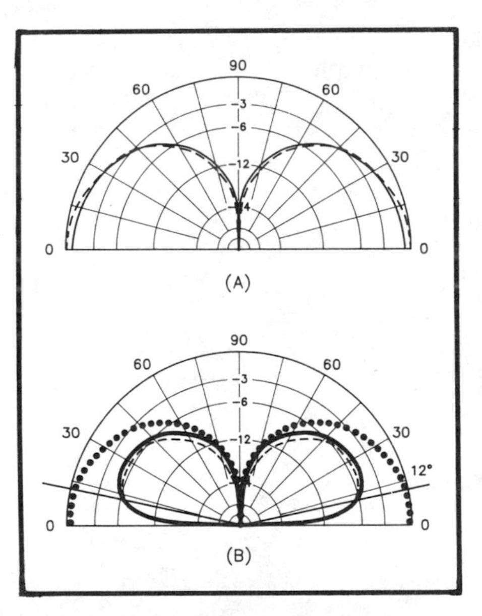

Fig 1—At A, the E-field patterns of a ¼-λ antenna and a short antenna differ only slightly over perfect earth. For equal radiated power, the similarity is also present over real earth, as shown at B. Over real earth, both patterns are down by approximately 6 dB from the perfect-earth field value at the psuedo-Brewster angle of 12°—poor earth conditions typical of the suburban backyard surrounded by houses. *(Patterns calculated with MN).*

Solid lines—Patterns for a short vertical antenna.
Broken lines—Patterns for a ¼-λ 160-m vertical antenna.
Dotted lines—Pattern for a ¼-λ antenna over perfect earth.

Antenna 1—The ¼-λ Vertical

By definition, a ¼-λ antenna has an electrical length (height) of 90°. R_r by Eq 2 is the conventional 36.6 ohms. In series with the assumed R_g of 15 ohms, a feed-point resistance R_b of 51.6 ohms results. Because efficiency is the ratio of R_r to the total feed-point resistance including all losses (Eq 4), an efficiency of 71% is indicated. These data are listed in Fig 2 under Antenna 1.

Antenna 2—The Short Base-Loaded Vertical

A 40-foot vertical has an angular height of 27.8° (Eq 3) and R_r of 2.24 ohms (Eq 2) at 1.9 MHz. Being short of ¼-λ resonance, it exhibits capacitive reactance as part of its feed-point impedance. The capacitive reactance can be computed by viewing the antenna from the base as a transmission line terminated in an open circuit. The characteristic impedance of this "line" for the 40-foot height and 1.6-inch diameter is calculated as 365 ohms from Eq 5. Its input reactance, using Eq 6, will be $-j692$ ohms. This capacitive reactance can be canceled by a base-loading coil with an inductive reactance of $+j692$ ohms. This requires an inductance of 58 μH (Eq 7). If the 58-μH coil has a Q of say, 300, its loss resistance, R_c, will be 2.3 ohms (Eq 8), which will appear at the base in series with the R_g of 15 ohms and the R_r of 2.24 ohms for a total R_b of 19.54 ohms. Efficiency (Eq 4) is 11.5%. Note the triangular current distribution. The loss in any required L network to match these antennas to a 50-ohm coax feed line is assumed negligible.

We can expect the signals from this 40-foot base-loaded antenna to be down about 8 dB (Eq 9) from a full ¼-λ antenna in the *same location* with the *same ground-loss resistance*.[1] Of course, very few ¼-λ towers are so situated.

The ground loss is the thing which we must endure since we can do little to change it in limited space for a radial system. The

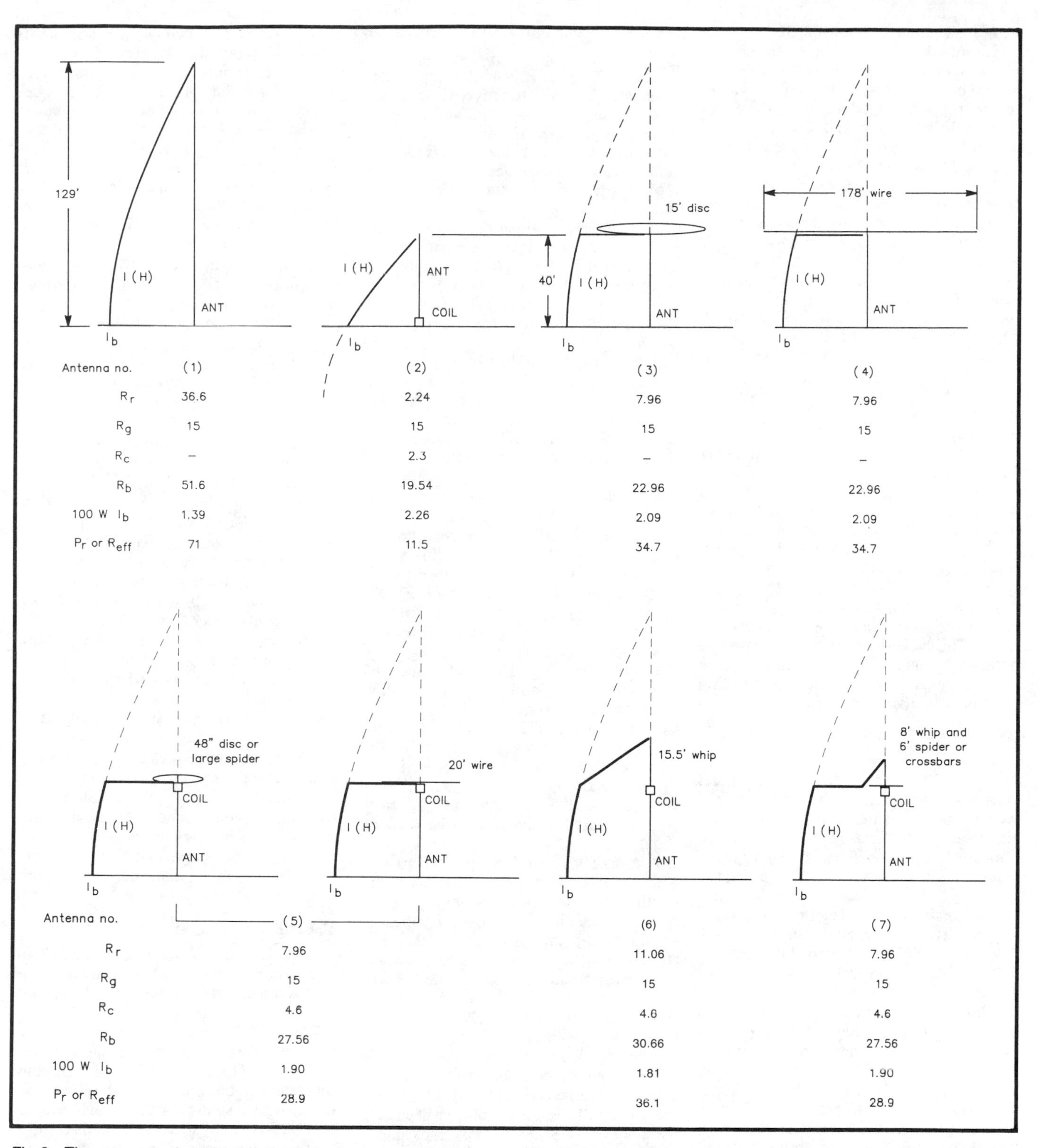

Fig 2—The antennas described in the text are shown with their current distributions, radiation resistances R_r, assumed ground-loss resistances R_g, coil loss resistances R_c (if any), total base input resistances at resonance R_b, base currents I_b for 100 W input to the antenna, and efficiency in percent. The efficiency percentage is the same as the radiated power in watts.

other factor, R_r, however, is something we can change.

Antenna 3—Capacitive Top Loading

At higher frequencies, a capacitive top hat is often used to increase the radiation resistance of a short antenna and bring it to ¼-λ resonance. This would eliminate the need for the base loading coil with its 2.3 ohms of loss resistance. The radiation resistance of the base section of a 40-foot top-loaded vertical can be calculated by Eq 1, but again a simpler equation (Eq 10) yields values sufficiently accurate for our purposes. For this antenna an R_r of 7.96 ohms results. This increases efficiency to 34.7% for a gain of 4.8 dB over the base-loaded system.

Shown in Fig 2 as Antenna 3, the improvement can also be seen as the change in current distribution compared with that of Antenna 2. The current times the length is the area under the current curve along the antenna. Power radiation is proportional to the square of this area, so the power radiated by Antenna 3 is 3 times that of Antenna 2 for the same input power—a gain of about 4.8 dB.

The top loading seemingly replaces the antenna portion and current distribution shown in broken lines. Therefore, some interpret top loading as increasing the effective height of an antenna. The top-loaded antenna, however, will not have the R_r implied by its ¼-λ resonance.

The 40-foot mast, viewed figuratively from the top as a 365-ohm transmission line, appears as a short-circuited line. The bottom end is terminated in an impedance that is very low compared to its characteristic impedance. Using Eq 11, a short-circuited line that is 27.8° long with a Z_0 of 365 ohms has an input impedance of $j192$ ohms. A capacitive reactance of 192 ohms would bring it to resonance. The corresponding capacitance is 436 pF by Eq 12.

A solid thin disc provides just about the highest capacitance available for a given area. Depending on the method used to calculate the disc size, it will be approximately 15 feet in diameter. The customary equation used to calculate the capacitance of a disc (Eq 13) is not applicable to such a large disc only 40 feet above the ground. At any rate, a purely capacitive top load seems impractical for such a short antenna at this frequency.

An intermediate arrangement could be a reasonably sized capacitive hat on Antenna 2. This would somewhat increase the antenna R_r and somewhat decrease the base inductance required to resonate it, resulting in improved efficiency over Antenna 2.

Antenna 4—The Wire Flattop

A horizontal wire flattop can be used to bring the 40-foot mast to ¼-λ resonance. As shown for Antenna 4, the center of the wire is connected to the top of the mast. Equal currents flowing outward cancel almost all radiation from the flattop. To bring a short antenna to ¼-λ resonance, the wire length should be about twice the length of the missing angular length. In this case, at 1.9 MHz, that length is approximately 2 × (90° − 27.8°) = 124° or 178 feet. This is 89 feet of wire on each side connected to the top of the 40-foot mast. But 178-foot flattops are not usually possible in this environment. Drooping the flattop wires eliminates the two end-supporting masts, but does not reduce the space required by a significant amount. It produces a downward-flowing current component which is in opposition to the current in the mast, slightly reducing R_r. A long but somewhat narrow space may accommodate such a flattop, and efficiency of 34.7% would be effected.

Good quality insulators must be used at the wire ends because of the rather high voltages present. Strain insulators ordinarily used in guy wires will not do.

The wire flattop is a cousin to the inverted L, which is really not a top-loaded vertical but an antenna that is a combination of vertical and horizontal elements.

Antenna 5—The Inductive-Capacitive Top Load

The 7.96-ohm R_r of the top-loaded antenna still looks attractive. If an inductive reactance is placed in series with a capacitive reactance, it reduces the effective capacitive reactance. Inductive reactance can make a small capacitance (high reactance) look like a large capacitance (lower reactance), as shown by Eq 14.

If a more practical hat size, say 48 inches in diameter, is postulated, its capacitance as calculated from Eq 13 is about 43 pF. Its reactance is $-j1946$ ohms (Eq 15). As previously calculated, the inductive reactance of the mast as seen from the top is $+j192$ ohms, leaving 1946 − 192 = 1754 ohms to be supplied by the inductor. From Eq 16, this is 147 μH, a feasible coil. Assuming that a Q of 300 is reasonable, then the coil loss resistance given by Eq 8 is 5.85 ohms.

This 5.85 ohms of resistance is located at the top of the mast. Because the current is sinusoidally distributed, the current at an angular distance from the current loop (at the base in a ¼-λ resonant antenna) is the loop current multiplied by the cosine of the angular distance from the loop. Since $P = I^2R$, we can "refer the resistance to the loop" by multiplying it by $\cos^2\theta$, Eq 17. Since the cosine of θ decreases with antenna height, taller antennas reflect less of the coil resistance to the base.

For the 27.8° mast, Eq 17 refers the 5.85 ohms to the feed point as 4.6 ohms, a loss resistance that becomes part of R_b. With R_r of 7.96 and R_b of 7.96 + 4.6 + 15 = 27.56 ohms, efficiency by Eq 5 is 28.9%.

A gain of 4.0 dB has been achieved over the base-loaded antenna, Antenna 2. Don't sniff at 4 dB! Some hams put up two-element phased arrays to achieve 4 dB of gain.

A 20-foot or so wire flattop of no. 10 wire with its center connected to the top of the loading coil could serve the same purpose as the 48-inch disc. Such an arrangement is essentially the wire flattop of Antenna 4 shortened by the action of the loading coil.

Wire lengths intermediate to the 178 feet of Antenna 4 and the 20 feet above could be resonated by progressively larger inductors, with space considerations or coil power loss determining the length of the wire. The same precautions regarding insulators apply. Tuning could be accomplished by pruning the wire length. Sloping the wire will change the tuning, because it affects the capacitance to ground.

Antenna 6—Inductive-Capacitive Top Loads With Whips

Another common technique is to use a whip above a loading inductor in a vertical antenna, such as in center loading or above-center loading. A whip as seen from the inductor can be treated as an open-circuited transmission line.

Assume a 15.5-foot whip with an average diameter of 0.562 inch (1 inch tapering to 1/8 inch). Anything longer seems rather difficult to support without guying somewhere above the coil. This is difficult because very high voltages are present on the top-loading coil and all parts of capacitive structures above the coil.

Eq 3 gives the whip angular length as 10.78°. The Z_0 is 371 ohms by Eq 5. The mast provides 192 ohms of inductive reactance as previously calculated. Eqs 6, 14 and 7 yield an inductance of 147 μH.

An antenna with a vertical section above the coil is called "segmented," or loosely, "center-loaded" (because the coil is not necessarily in the actual center). The top segment or whip increases the R_r, Eq 18 applies (see the "Appendix") and yields an R_r of 11.06 W. Using Eq 4, the efficiency is 36.1%. From Eq 9, gain over the base-loaded Antenna 2 is 5 dB, or about 1 dB over Antenna 5. The rather large whip, however, has extended the overall height to 55.5 feet, with attendant wind load and mechanical problems. A shorter whip requires a larger inductance and hence increases coil loss, which, for significantly shorter whips, may become intolerable. The same R_r and efficiency could be obtained from Antenna 5 by increasing its mast length from 40 feet to 47.9 feet.

The question often arises as to how much power is radiated by the top section. This 15.5-foot whip above the 40-foot base section increases R_r over R_r of the disc-loaded Antenna 2 by 28%, but analysis of the current distribution indicates that somewhat less than 3.6% of the power is radiated by the whip.

Under the assumptions made, the inductive-capacitive top-loaded Antennas 5 and 6 should provide signals about 4 and 3 dB less, respectively, than that from a quarter-wave antenna in the same location with the same ground-loss resistance.

Capacitive Structures

The capacitive structures described represent just about the largest disc or whip that is manageable, although I once heard a Texas station (where else?) with a hat made of 24-foot crossed sections of irrigation tubing with wires connecting the ends and the midpoints. Whips of generous diameter can be combined with rather long crossbars or spiders with enough legs to approach the capacitance of a disc for a more manageable and practical capacitive structure.

Whips and spiders or crossbars seem to add their capacitance fairly well, but no combination is equal to the sum of the capacitance of its parts. The larger the hat just above the coil, the less the whip contributes.

Antenna 7 of Fig 2 illustrates the current distribution of a combined capacitive hat and whip antenna. The combination has an effective combined capacitance of 43 pF, requiring the same 147 μH of inductive reactance. Its characteristics are similar to those of Antennas 5 and 6.

Bandwidth

The inductive-capacitive top-loaded antenna described will exhibit bandwidths of approximately 20 kHz between the 2:1 SWR points. If other than fixed-frequency operation is planned, then it is best to design and trim for a natural resonance at 2 MHz and use a small base coil to move around the band. A variable inductor of about 30 μH should provide for tuning down from 2 MHz to 1.8 MHz. I use a remotely switched motor-driven inductor.

Reduction of loss will decrease the bandwidth. The limiting case of essentially no ground loss would probably yield a bandwidth of about 10 kHz, but then other problems become quite severe.

Effect of Coil Loss

Larger, higher capacitance structures reduce the inductance requirements. Heroic structures are required to get the required inductance down to the 80-μH range.

Because of the swamping effect of the assumed 15-ohm R_g, loss in a top-loading coil has surprisingly little effect on efficiency over a rather large loss range. With 100 W to the antenna, loss in the 147-μH coil with a Q of 300 would be 13 W, and the peak potential across the coil about 4000 volts. This is quite acceptable for coils of modest size and construction.

Operation of antennas such as Antenna 7 at the legal limit of 1500 W to the antenna would produce power loss in the coil of 250 W and peak coil voltage of 20,000. While the 250 W would average out to rather modest levels for SSB operation, the peak voltage would still exist. Under "key-down" or AM operation, the 250-watt level would be catastrophic to coils of the usual size and construction. For legal-limit input, the top-loading structure will have to be such as to require less inductance, or the coil will have to be designed for a substantially higher Q, or both.

For example, say a coil could safely dissipate a power on the order of 150 W. This is probably possible for a good-sized coil in an open-air environment. If so, then coils ranging from 80 μH with a Q of 300 (requiring very large capacitive structures) to 150 μH with a Q of 600 (requiring the reasonable structures of Antenna 7) will survive.

The alternative is to ensure that the duration of such power input never exceeds some brief time well within the thermal time constant of the coil—a risky procedure at best. Stories of burning or melting top-loading coils are not uncommon.

Effects of Ground-Loss Reduction

The efficiency of these short top coil-loaded antennas is improved with reduction of ground loss. An extensive radial system and earth of good quality produces a loss resistance of about 2 ohms. Over such a ground system, Antenna 7, with a 147-μH coil of Q = 300, will have an efficiency of almost 54%, for a signal increase of 2.5 dB. Although hardly worth the effort, this is possible because at a 100-watt power level the coil loss would be about 32 W.

Operation of this same antenna over such a ground system at 1500 W would lead to a coil loss of 475 W and 20,000 volts across the coil. Even a 147-μH coil with a Q of 600, which is quite hard to achieve in practice, would have to dissipate 300 W. An 80-μH coil with its attendant capacitive structure problem would dissipate 150 W. In summary, high-power operation of these short inductive-capacitive 160-meter antennas over good ground systems is limited by coil heat dissipation and voltages. The base-loaded vertical, with its loading coil at ground level, makes exotic coils possible, but the better solution is Antenna 4, the T with the horizontal-wire top load. Antenna 5, using a wire flattop of intermediate length, could reduce loading-coil requirements to a range of permissible loss, with space requirements being the trade-off.

The best article on ground radial systems that I have seen in amateur literature is one by Brian Edward, N2MF.[2] His Fig 7 is particularly applicable to ground systems of the kind likely to exist under the space limitations in which short antennas are often situated.

Vertical antennas are sometimes "low tuned" by top loading to raise the current loop farther above ground to reduce ground losses. In a short antenna, the angular distance just is not there to allow much current difference, and the required larger and hence more lossy inductor incurs additional coil loss that may exceed the ground-loss reduction.

The antennas discussed are all assumed to be base fed. Grounded towers can be top loaded to facilitate shunt feed or slant-wire feed. Short folded monopoles, folded umbrellas etc on grounded masts can also be top loaded. Although these feed methods yield a different (usually higher) feed-point impedance, this transformation of impedances does not affect the ground currents or ground losses of the antenna. Such currents and losses will be the same as they would be if the tower was isolated from ground at the base and fed at that point.

Conclusions

Short inductive-capacitive top-loaded antennas are suitable for operation under the conditions assumed. Coil loss is the compromise to space considerations, and the limiting factor for high power operation.

Coil-loss problems decrease with increased mast height or the availability of horizontal space for such as horizontal-wire top-capacitive structures. Reduction of coil loss will improve efficiency to some extent and permit operation at higher power levels.

Very low-loss ground systems, although improving efficiency for low-power operation, place prohibitive requirements on the loading coils for even large whip and spider capacitive structures. Survival of conventional loading coils with high power input is most probably because of very high ground loss.

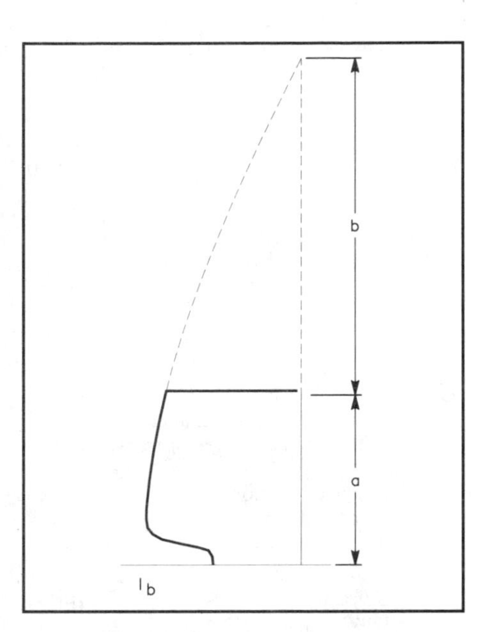

Fig 3—Showing how lengths a and b are defined for Eq 1 of the Appendix. The total length (height), a plus b, equals 90°.

Appendix

$$R_r = \frac{30\ f(\text{height})}{\sin^2 G} \qquad \text{(Eq 1)}^3$$

where

$$f(\text{height}) = \sin^2 B \left[\frac{\sin 2A}{2A} - 1\right] - \frac{\cos 2G}{2}\left[S_1(4A)\right] + S_1(2A)(\cos 2G + 1) + \sin 2G\left[\frac{S_i(4A)}{2} - S_i(2A)\right]$$

$A = 2\pi a/\lambda$

$B = 2\pi b/\lambda$

a = length a, feet (see Fig 3)

b = length b, feet (see Fig 3)

λ = wavelength, $\frac{984}{f(\text{MHz})}$ feet

$G = A + B$

$$S_1(x) = \int_0^x \frac{1-\cos x}{x}\,dx = \frac{x^2}{2\times 2!} - \frac{x^4}{4\times 4!} + \frac{x^6}{6\times 6!} \cdots \frac{x^n}{n\times n!}$$

$$S_i(x) = \int_0^x \frac{\sin x}{x}\,dx = x - \frac{x^3}{3\times 3!} + \frac{x^5}{5\times 5!} \cdots \frac{x^n}{n\times n!}$$

S_1 and S_i must be to about six-place accuracy for short antennas

$$R_r = 36.6 \frac{(1 - \cos H)^2}{\sin^2 H} \quad \text{(Eq 2)}^4$$

where

R_r = radiation resistance of simple monopole, ohms
H = angular height of simple monopole

$$H = 0.366hf \quad \text{(Eq 3)}$$

where

H = angular length, degrees; Note: 360/984 = 0.366
h = length, feet
f = frequency, MHz

$$Eff = 100 \frac{R_r}{R_b} \quad \text{(Eq 4)}$$

where

Eff = efficiency, %
R_r = radiation resistance, ohms
R_b = total feed-point resistance, ohms

$$Z_0 = 60\left[\ln\left(\frac{48\ h}{d}\right) - 1\right] \quad \text{(Eq 5)}$$

where

Z_0 = characteristic impedance of vertical monopole considered as a transmission line, ohms
h = height of monopole, feet
d = average diameter of monopole, inches
ln = natural logarithm

$$Z_{oc} = \frac{-jZ_o}{\tan\theta} \quad \text{(Eq 6)}$$

where

Z_{oc} = input impedance of open-circuited line, ohms
Z_0 = characteristic impedance of line, ohms
θ = angular length of line
j = the complex operator

$$L = \frac{X_L}{2\pi f} \quad \text{(Eq 7)}$$

where

L = inductance, μH
X_L = inductive reactance, ohms
f = frequency, MHz

$$R_c = \frac{X_L}{Q} \quad \text{(Eq 8)}$$

where

R_c = resistance of coil, ohms
X_L = inductive reactance of coil, ohms
Q = quality factor of coil

$$dB = 10 \log_{10}\left(\frac{Eff_1}{Eff_2}\right) \quad \text{(Eq 9)}$$

where

dB = gain, decibels
Eff_1 and Eff_2 = efficiencies being compared

$$R_r = 36.6 \sin^2 H \quad \text{(Eq 10)}^5$$

where

R_r = radiation resistance of the base section of a top-loaded antenna at ¼-λ resonance
H = angular height of base section

$$Z_{sc} = jZ_0 \tan\theta \quad \text{(Eq 11)}$$

where

Z_{sc} = input impedance of short-circuited transmission line
Z_0 = characteristic impedance of line, ohms
j = the complex operator
θ = angular length of line

$$C = \frac{10^6}{2\pi f X_c} \quad \text{(Eq 12)}$$

where

C = capacitance, pF
f = frequency, MHz
X_C = capacitive reactance, ohms

$$C = 0.8992d \quad \text{(Eq 13)}$$

where

C = capacitance, pF
d = diameter, inches

$$X = X_C - X_L \quad \text{(Eq 14)}$$

where

X = resulting reactance
X_C = capacitive reactance, ohms
X_L = inductive reactance, ohms

$$X_C = \frac{10^6}{2\pi f C} \quad \text{(Eq 15)}$$

where

X_C = capacitive reactance, ohms
f = frequency, MHz
C = capacitance, pF

$$L = \frac{X_L}{2\pi f} \quad \text{(Eq 16)}$$

where

L = inductance, μH
X_L = inductive reactance, ohms
f = frequency, MHz

$$R_{Loop} = R_\theta \cos^2\theta \quad \text{(Eq 17)}$$

where

R_{Loop} = resistance at θ transferred to current loop
R_θ = resistance at θ from current loop
θ = angular distance between resistance to be transferred and current loop

$$R_r = 36.6\left[\sin H_1 + \left(\frac{\cos H_1}{\sin H_2}\right)(1 - \cos H_2)\right]^2 \quad \text{(Eq 18)}^6$$

where

R_r = radiation resistance of a segmented antenna at quarter-wave resonance.
H_1 = angular length of base section.
H_2 = angular length of top section.

Notes

[1] [**EDITOR'S NOTE**: As Fig 2 shows, with equal power applied, more current flows at the base in the shorter, base-loaded element than in the full-size, ¼-λ element. Intuitively, it may then seem that this higher base current might yield a field-strength increase (gain) to offset some of the resistive losses, and therefore the author's figure of "8 dB down" may appear to require modification. However, not only the current *amplitude* but also the *current distribution* in the conductor (as indicated in Fig 2) is a factor in determining far-field signal strength. The 8-dB difference can be verified with antenna analysis programs using method of moments calculations, such as NEC, MININEC and MN. Other antenna configurations evaluated in this article can be similarly verified.]

[2] B. Edward, "Radial Systems for Ground Mounted Vertical Antennas," in Chapter 6.

[3] Adapted from G. H. Brown's thesis, "A Theoretical and Experimental Investigation of the Resistances of Radio Transmitting Antennas," Univ. of Wisconsin, 1933, citing van der Pol and R. Bechmann, *Jahrbuch O. Drahtl Telegr*, 13, 217, 1918.

[4] From B. Byron (W7DHD), "Short Vertical Antennas for the Low Bands," *Ham Radio*; Part 1, May 1983, pp 36-40, and Part 2, Jun 1983, pp 17-20.

[5] See Note 4.

[6] The author's extension to the general case from the center-loaded derivation by J. Evans, NV1W, by the method of W. Byron (see Note 4) in private communication.

from *The ARRL Antenna Compendium, Vol 4*

Short Coil-Loaded HF Mobile Antennas: An Update and Calculated Radiation Patterns

By John S. Belrose, VE2CV, ARRL TA
17 Tadoussac Drive
Aylmer, QC
J9J 1G1, Canada

There is more to consider than appearance and convenience when installing your HF mobile antenna, since the frame and the body of the vehicle are a part of the radiating system.

Introduction

It has been known for some time that the efficiency of an electrically short whip is improved when the tuning coil is moved from the monopole base (*base fed*) and located in series with the monopole itself. The coil is then called a *loading coil*. The earliest paper on inductively loaded HF mobile antennas is by Belrose,[1] who analyzed the antenna as an opened-out transmission line. Although this analysis is only approximate, the trends are correct. The results obtained agree rather well with the more rigorous analysis by Hansen,[2,3] who analyzed the inductively loaded antenna by numerical moment-method techniques. Both authors showed that maximum efficiency for a monopole of height h occurs at a load point approximately 0.4 h from the feed.

The author's early (simple) analysis assumed a linear current distribution on the antenna, an assumption that limits the applicability of the method. Later this restriction was removed (see Note 4 and Annex A) by assuming a sinusoidal current distribution on the antenna. The modified analysis can be used to calculate the radiation resistance with reasonable accuracy for heights up to about 0.2 λ.

However, the analyses described above neglected the effect of the mounting structure on antenna performance. From personal experience, the author has recognized that received signal strengths using a rear-bumper-mounted HF mobile whip on the 20 meter band can be up to two S-units (about 10 dB) stronger when the vehicle is heading in the direction of propagation. It is clear that currents on the frame and body of the vehicle have a profound influence on the performance of the antenna.

In this article I summarize results of recent measurements, and of a detailed numerical modeling study employing *MININEC*,[5] and *NEC-2*,[6] for an HF center-loaded mobile antenna bumper mounted on a GMC Jimmy truck. The study begins with the calculated performance for a ground-mounted HF center-loaded monopole. These results can be compared with the earlier simple method of analysis.

Ground Mounted

Three types of electrically short monopoles are sketched in **Fig 1**. Here we are concerned with the inductively loaded monopole in Fig 1B. The monopole height is $h = h_1 + h_2$ and d_1 and d_2 are the average diameters of each section. These antenna dimensions can be measured in feet, inches, meters or millimeters. The author uses mm in his detailed numerical models.

The first case study is for a 110-inch center-loaded mobile monopole. Here, h = 2794 mm; $h_1 = h_2 = 1397$ mm; $d_1 = 5.55$ mm; and $d_2 = 15.9$ mm. The inductance L_0 is adjusted so that the base input impedance of the center-loaded whip is purely resistive; that is, the antenna is resonant. To do this by numerical modeling using *MININEC* (*ELNEC* or *MN* version) proceed as follows:

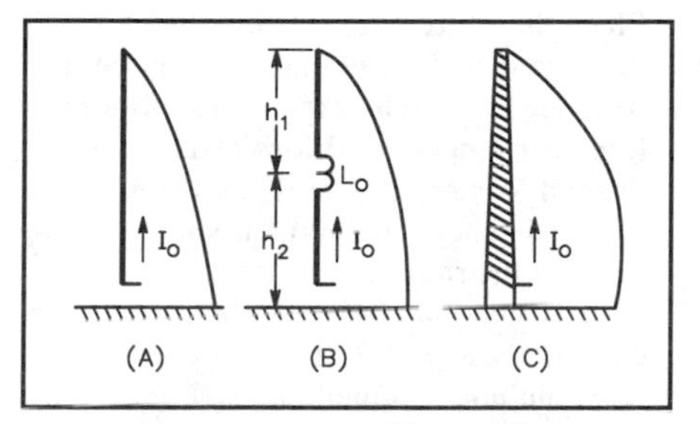

Fig 1—Electrically short monopoles: at A, base-loaded; at B, inductively loaded; and at C, continuously loaded.

Table 1

A Comparison between the Simple Analysis (Sinusoidal Current Distribution) and *ELNEC* for the Case Study Antenna Ground Mounted

	Frequency MHz	R_r Ω	X_c Ω
Simple Analysis	3.75	1.15	*j* 2922
Ground Mounted	7.15	4.22	*j* 1598
	14.15	15.40	*j* 807
ELNEC	3.75	1.20	*j* 3020
Ground Mounted	7.15	4.19	*j* 1520
	14.15	14.80	*j* 631

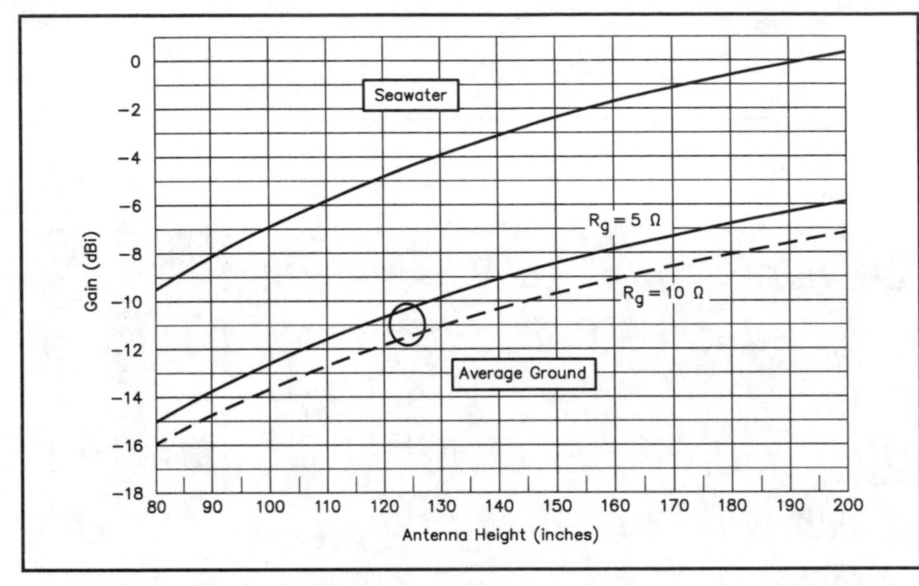

Fig 2—Gain in dBi of a coil-loaded HF mobile antenna versus antenna height, where the antenna is over seawater and then over ground of average conductivity. The frequency is 3.75 MHz (80 meters), and $h_2/h = 0.4$.

1) Place the load and the source at the junction between h_1 and h_2, and set the load impedance to zero (zero reactance and zero resistance).
2) Calculate the source impedance. For our mobile antenna at 3.75 MHz, $X_s = -j$ 3055 Ω. This is the reactance to be canceled by the inductive reactance of the loading coil L_O.
3) Set the load reactance to the required value (+*j* 3055 Ω).
4) Place the source generator at the base of the antenna, and calculate the source impedance for zero loss resistance. Since all loss resistances (coil loss resistance R_c, ground loss resistance R_g and conductor loss resistance) are zero, the source resistance is the radiation resistance R_r referenced to the base of the antenna (for our case study $R_r = 1.2$ Ω).
5) Now include assumed losses. For the coil loss assume a Q-factor = 300; for the ground loss resistance assume for the present example $R_g = 12$ Ω, see below; and for the conductor loss enter the appropriate conductor material (aluminum assumed).
6) Calculate input impedance ($R_a = 26$ Ω). Radiation efficiency, antenna gain and bandwidth can be calculated.

The radiation efficiency of our 110-inch center-loaded whip at 3.75 MHz is

$$\eta = \frac{R_r}{R_r + R_c + R_g}$$

$$= \frac{R_r}{R_a} = \frac{1.2}{26} = 0.046 = 4.6\%$$

The calculated values compared with the analysis assuming a sinusoidal current distribution are tabulated in **Table 1**. For estimation of radiation efficiency and gain, a ground loss resistance of 12 Ω corresponds to measured values for an 80-meter HF mobile antenna on an automobile over average ground. For a ground-mounted antenna, this might correspond to a stake ground. For 7.15 and 14.15 MHz, the ground loss resistances for this case study are 9 Ω and 29 Ω; see below.

Effect of Real Ground on Antenna Performance

The gain of a monopole antenna over real ground is dependent on the conductivity of the ground beneath the antenna (which affects the radiation efficiency of the antenna), and in front of the antenna (which affects the vertical plane pattern of the antenna). When the conductivity of the ground is finite the antenna's performance is seriously affected for elevation angles less than 15°, angles which are comparable with the most probable arrival angle for HF skywaves from distant stations.

In the discussion above we implicitly assumed that the radiation efficiency for an electrically short monopole at a low height over real ground is determined by the radiation resistance divided by the total antenna resistance. The total antenna resistance is the sum of the radiation resistance (R_r), the ground loss resistance (R_g), the coil loss resistance (R_c) and the antenna's conductor loss resistance. Except for electrically small loop antennas, the conductor loss resistance is small with respect to the other resistances. The ground beneath the antenna introduces a loss resistance, which is a function of the frequency, the height of the antenna above the ground and the ground conductivity. The smaller the antenna the higher its capacitive reactance, the larger the tuning coil required for resonance, and the greater the loss associated with the coil-loss resistance. The gain of a short antenna therefore decreases as the frequency, height or length of the antenna decrease.

In **Fig 2** we have plotted the maximum skywave gain for an electrically short coil loaded antenna, computed by *ELNEC*, as a function of antenna height or length of the antenna. This is for antennas over seawater and over average ground (conductivity, $\sigma = 3$ mS/m, dielectric constant, $\varepsilon = 13$). The frequency is 3.75 MHz (80 meter), and $h_2/h = 0.4$. We have further assumed that $R_g = 0$ for the case of an antenna over seawater. Clearly, while the ground loss resistance is an important parameter, a change of ground loss resistance from 5 Ω to 10 Ω results in a decrease in gain by only 1 dB. A most important parameter is the finite conductivity of the ground in front of the antenna, which affects the ability of the antenna to launch skywaves.

Computer programs like *NEC* and *MININEC* express antenna gain with respect to an isotropic radiator (dBi). For electrically short vertically polarized antennas, a better reference might be either radiation efficiency (since power radiated can be calculated) or gain with respect to a well-grounded quarter-wave monopole. Both the antenna under study and the reference antenna are equally influenced by the conductivity of the ground in front of the antenna.

Vehicular Antenna—Ground Loss Resistance

Ground loss resistance is a function of vehicle size (electrical size in wavelengths), placement of antenna on vehicle (according to Brown[7]), and conductivity of the ground over which the vehicle is traveling. Newer, smaller vehicles with front-wheel drive do not have a complete frame. Undoubtedly they provide an inferior ground plane compared to older, full-size vehicles with rear-wheel drive. More than 40 years ago the author [previously referenced 1953 *QST* article] estimated that the ground loss resistance R_g for a 3.8 MHz whip on his 1940 Dodge was about 12.5 Ω.

The author has recently measured the ground loss resistance for a bumper-mounted Swan Model 45 HF mobile whip (one that had never been used) mounted on his 1992 GMC Jimmy truck. That is, the antenna was "new" and the truck was not rusted. A home-brew low-capacity bumper-mount base insulator was used. The antenna's input resistance at resonance was measured using modern instrumentation—a Hewlett-Packard Impedance/Gain/Phase Analyzer Model 4194A, configured to measure balanced impedance using an HP supplied balun. One terminal of the balanced input was connected to the antenna; the other terminal to the vehicle ground. The effect of the balun and the short lead length to reach the antenna were calibrated out, so the antenna's input impedance was measured with the vehicle in effect isolated from instrument and power ground.

To estimate ground loss resistance the antenna's input impedance at resonance was calculated using *ELNEC* and a wire grid model to simulate the vehicle (see below), for an assumed coil Q-factor of 300, and no ground loss resistance. The Swan Model 45 whip employs a large air-wound coil. The difference between the measured value of R_a (at resonance), which includes ground loss, and the calculated value assuming $R_g = 0$, gives the estimated value of ground loss resistance. With the whip's band selector switch set at the 3.8 MHz tap, the resonant frequency was 3.77 MHz, and the impedance at resonance $R_a = 22$ Ω. The calculated antenna resistance $R_r + R_c = 9.7$ Ω. Hence $R_g = 12.3$ Ω.

For 7.077 MHz (the antenna's resonant frequency on the 40-meter band) the antenna's resistance R_a measured is 16.4 Ω. The calculated antenna resistance $R_r + R_c = 7.55$ Ω. Hence $R_g = 8.8$ Ω.

For 14.79 MHz (the antenna's resonant frequency on the 20-meter tap position) the antenna's resistance R_a measured is 46.6 Ω. The calculated antenna resistance $R_r + R_c = 14.5$ Ω. Hence $R_g = 32$ Ω.

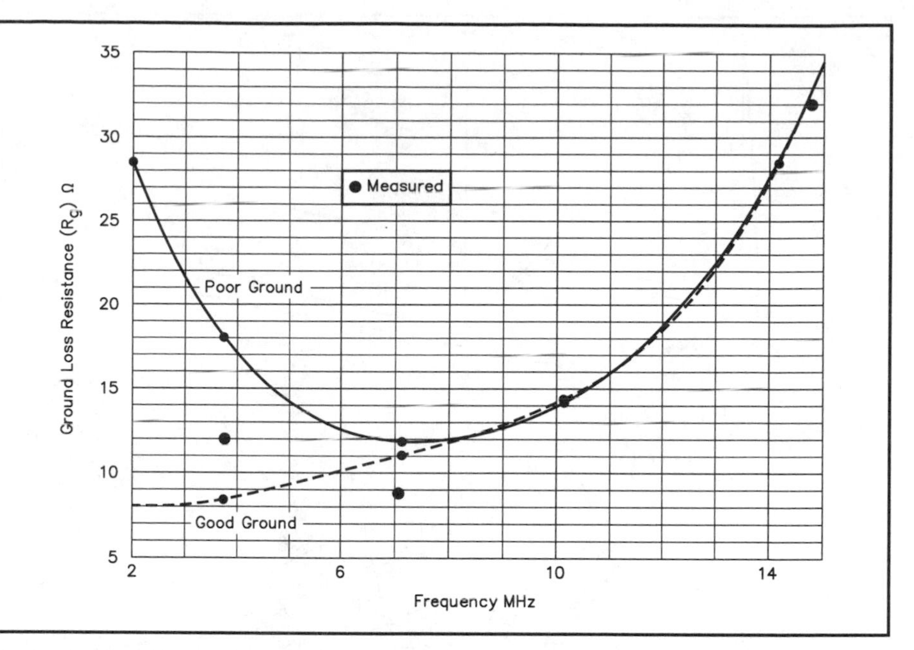

Fig 3—Ground loss resistance (R_g) measured (see text), and calculated using *NEC-2* for an electrically short HF mobile antenna on the basic frame of the vehicle, for two ground conductivities.

For a whip mounted on the basic frame of our vehicle (see below) we have calculated the ground loss resistance using *NEC/Wires 1.5* for good ($\sigma = 5$ mS/m, $\varepsilon = 13$) and for poor ($\sigma = 1$ mS/m, $\varepsilon = 5$) ground. The results of this analysis, and the measured values are plotted in **Fig 3**. This figure shows two interesting features: (1) There is a rather good agreement between the measured and calculated values for the ground loss resistance; and (2) The type of ground has a strong influence on the ground loss resistance for frequencies below about 7 MHz. For frequencies above about 7 MHz, the type of ground (good or poor) has little or no effect.

Vehicular Antenna—Radiation Pattern

First we will consider an inductively loaded antenna fed against the frame of the vehicle (see **Fig 4A**). The reason for doing this is two-fold. First, no bumper bracket is needed—a bumper bracket is a very short wire and this may give trouble with the *MININEC* model. Second, the return currents that flow on the basic frame are orthogonal to the current on the antenna. Radiated fields due to these return currents therefore cannot destructively interfere with radiation due to current on the whip. We will further discuss this below.

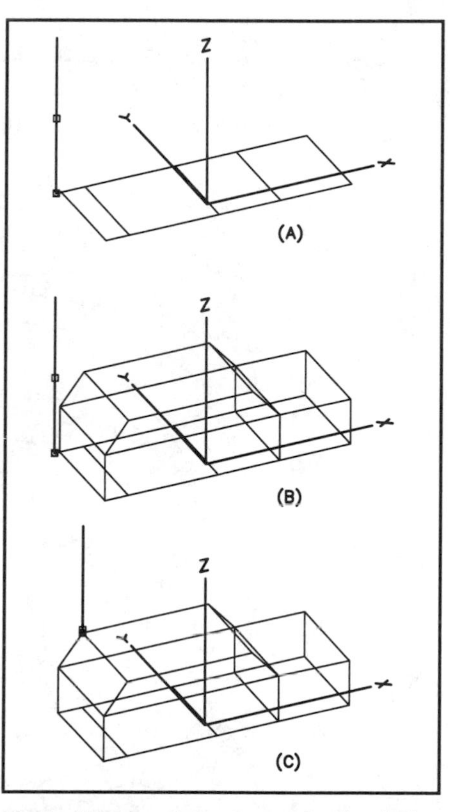

Fig 4—Wire models of a center-loaded HF mobile antenna: at A, rear left-bumper mount on the basic frame of a motor vehicle; at B, on a GMC Jimmy truck; and at C, on the left-rear corner of the roof (whip tip height the same as for B).

The currents on the wire model are shown in **Fig 5A**. The radiation patterns, elevation pattern for azimuth angle $\varphi = 337°$ (the azimuth of maximum gain), and the principle plane azimuth pattern (elevation angle of 31° for maximum gain) for 3.75 MHz are

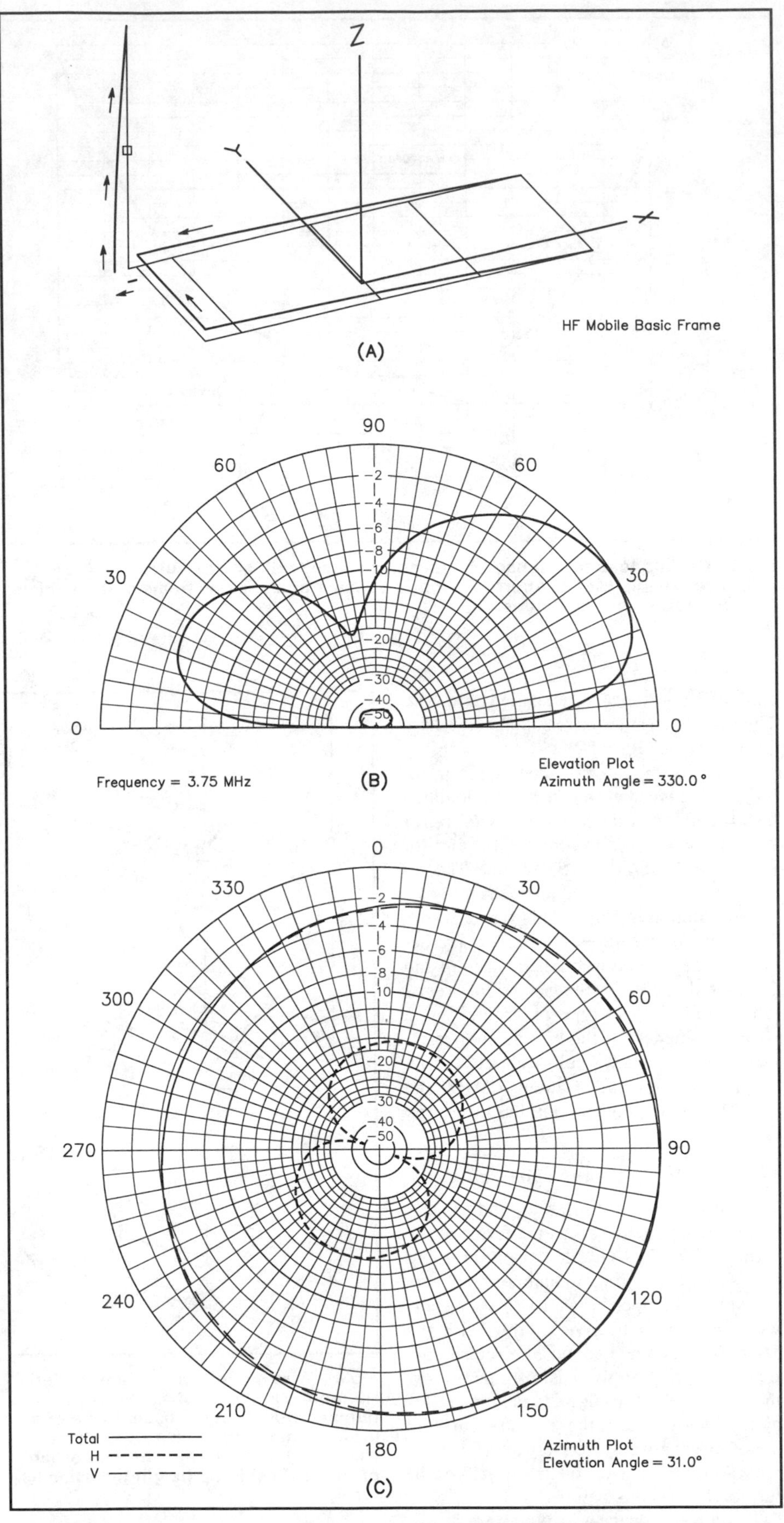

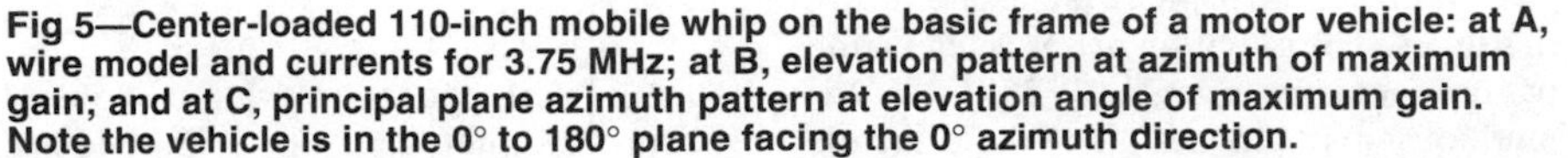
Fig 5—Center-loaded 110-inch mobile whip on the basic frame of a motor vehicle: at A, wire model and currents for 3.75 MHz; at B, elevation pattern at azimuth of maximum gain; and at C, principal plane azimuth pattern at elevation angle of maximum gain. Note the vehicle is in the 0° to 180° plane facing the 0° azimuth direction.

given in Fig 5B and 5C. Notice that the maximum gain is in the direction of the long axis of the vehicular frame. For a left-rear-corner-mounted antenna, maximum gain is in the diagonally opposite direction.

We calculated the radiation resistance (for loss resistances zero) $R_r = 1.3\ \Omega$, the loading coil for resonance and the gain for a loss resistance $R_g = 12\ \Omega$, with and without a short bumper bracket. A bumper bracket is needed for the Jimmy vehicle model; see below. The reactance of the loading coil for resonance and the gain agreed almost exactly, thus validating the model with a bumper bracket.

Let me comment on the model, for those radio amateurs interested in modeling. The segment length on all wires simulating the vehicle and the frame of the vehicle are made approximately the same length (number of segments proportional to wire length). Four segments are used on the bracket on which the whip is mounted, and the segment tapering option (*ELNEC* program) was used, tapering to this segment length at the source and load ends of the antenna wires.

Now let us look at a bumper-mounted antenna on a Jimmy truck; for the wire-grid model see Fig 4B. Here, return currents on the conductors of our wire-grid model of the vehicle can be oppositely directed to the current on the antenna, see **Fig 6A**. The resultant radiation field associated with these currents is in part canceled. This results in a lower effective value of the radiation resistance. For this antenna at 3.75 MHz, $R_r = 0.6\ \Omega$. Compare with the value for the antenna mounted on the frame, where $R_r = 1.3\ \Omega$. The whip and the vehicle body are closely coupled—a change of current on the antenna is reflected by a change of current on the body of the vehicle in a complicated way, and the wire-grid model may not exactly simulate the metal surface of the vehicle. Nevertheless, the author considers the results obtained to be reasonable, and the radiation patterns (Fig 6B and 6C) are in accord with expectation. Compare these with Fig 5B and 5C.

Let us now consider further the reduced value of the radiation resistance due to the oppositely directed return currents on the body of the vehicle. The calculated radiation resistance for an antenna having the same tip height, mounted on the left-rear corner of the roof, is $R_r = 0.53\ \Omega$. See Fig 4C. This is only marginally less than the value for the bumper-mounted whip. Indeed, fields due to current flow on the lower part of the mobile antenna and the vehicle almost cancel.

Finally, we have modeled the vehicle with a roof-mounted whip, which is not a very practical installation for passing under low bridges and beneath overhanging trees.

This lifts the antenna and reduces to a minimum the effect of oppositely directed return currents. The antenna is more or less symmetrical with respect to the support structure. The calculated radiation patterns for 3.75 MHz are given in **Fig 7**.

Radiation Pattern for 40 and 20 meters

For the amateur interested in mobile operating at high frequencies, the patterns in Fig 4B for 40 and 20 meters are plotted on **Fig 8** and **Fig 9**. The tuning inductances are j 1559 Ω and j 647 Ω respectively. Clearly, the basic azimuth pattern found for the 75/80-meter band is retained, and the efficiency and directivity increase as frequency increases.

Predicted Gain

The predicted gains for 80, 40 and 20 meters for a 110-inch whip on the GMC Jimmy truck are –13.5 dB, –6.2 dB and –3.8 dB with respect to a well-grounded quarter-wave monopole respectively. The front-to-back ratio is about 9 dB at 14.15 MHz, which is in agreement with operational experience.

HF Mobile for NVIS

The traditional HF antenna used with land vehicles is a vertical whip, which produces little radiation straight up, making it poorly suited for near vertical incidence skywave (NVIS) communications, a typical requirement for mobile operators using the 80 and 40-meter bands. For 20 meters and up a low launch angle is desired. The approach to overcome this problem for military NVIS tactical communications involves the use of a tilted whip. Wallace[8] described the use of a whip-tilt adapter that raised the whip 2 feet over the vehicle's top, with the top part of the whip pointed forward. He claims that this works well for NVIS communications. The author modeled such an antenna system, see **Fig 10A**. There does not seem to be much gained (see Fig 10B) over a bumper-mounted mobile whip. Even though the antenna is much longer (20 feet) the maximum gain according to *ELNEC* is less, and the gain at high elevation angles is not improved. But the pattern is omnidirectional (Fig 10C). Perhaps it is this difference that leads military communicators to believe that the tilted whip resolved the NVIS communications problem.

If the whip can be tilted, an alternative arrangement is to direct the whip rearward, at an angle to the ground, although this is hardly a practical solution for a vehicle in motion. This tilted-back arrangement provides high-angle radiation, since the whip and the rubber-tired vehicle radiate somewhat like a funny-looking dipole, the vehicle being one "arm" of the dipole. However, it should be noted that military whips

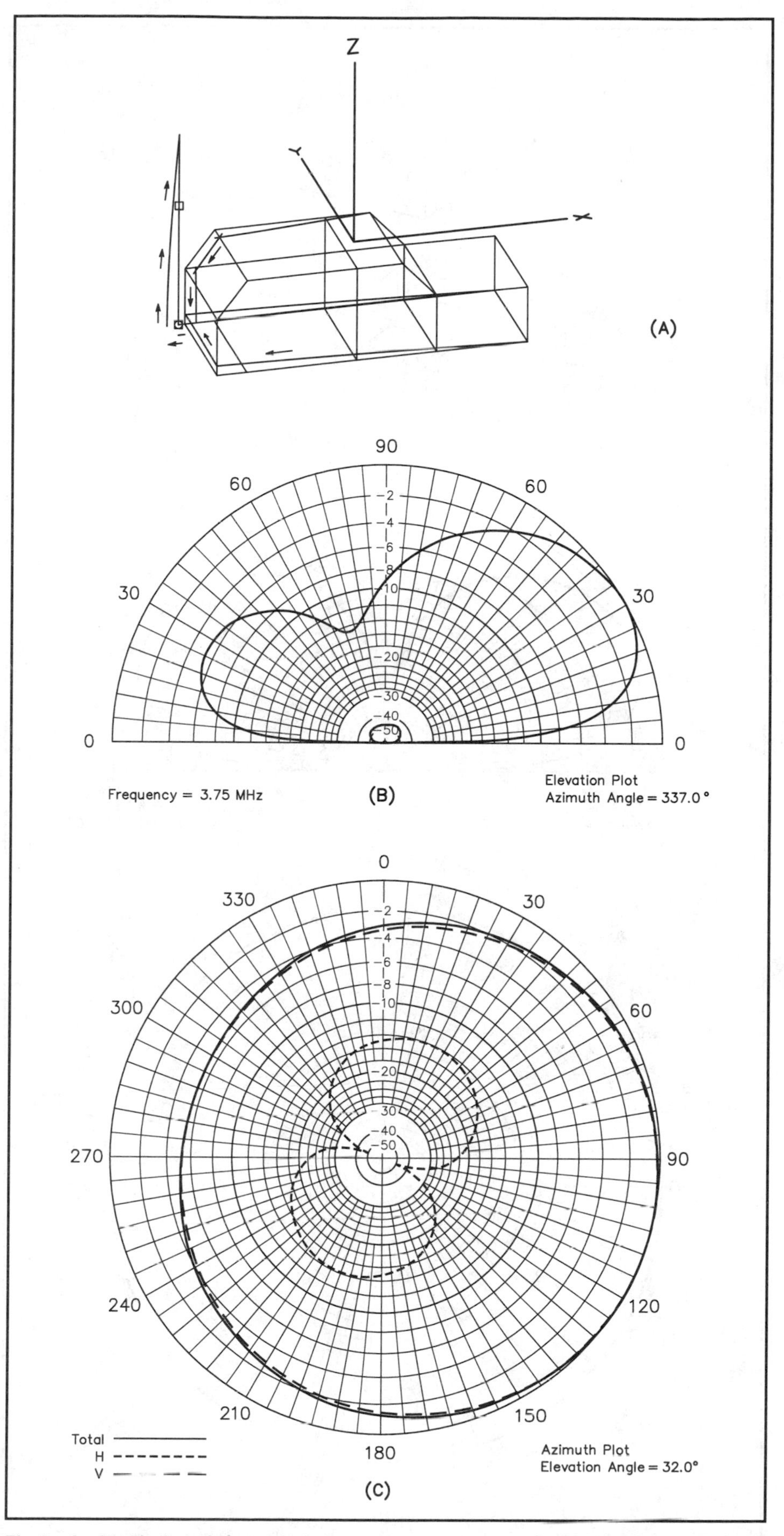

Fig 6—As for Fig 5, but the whip is bumper-mounted on a GMC Jimmy truck (Fig 4B), frequency 3.75 MHz.

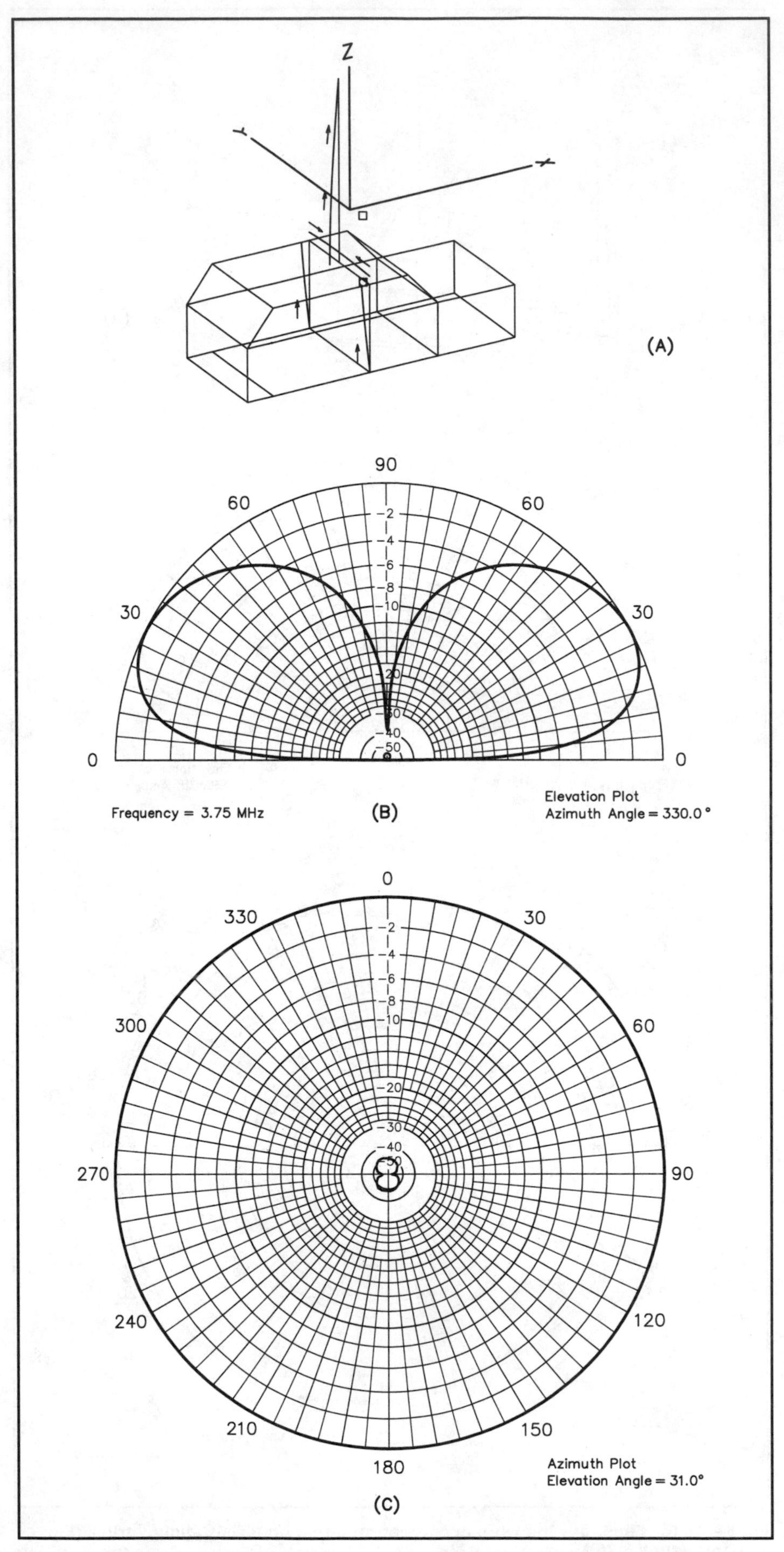

Fig 7—Roof-mounted 110-inch center-loaded whip, frequency 3.75 MHz.

are 16 feet long, which is quite a different length compared with the 9 foot (110-inch) whip described here. In fact, the military whips can be extended all the way to 32 feet. It is no wonder that only the lower half of the whip is used, and the end of the whip is tied down when traveling.

Practical Considerations: Tuning/Matching and Installing HF Mobile Antennas

HF mobiling continues to be interesting to the HF Amateur Radio community, as is perhaps evident by two articles published in *QST* in 1993. Jeff Gold, AC4HF,[9] briefly described his experience with a number of commercially available HF mobile antennas. Roger Burch, WF4N,[10] describes how he made his own HF mobile antenna and mount. A number of the author's fellow amateurs are duplicating D. K. Johnson, W6AAQ's design (see below).

In his own 1953 *QST* article, the author addressed the subject of tuning and matching the HF mobile antenna. He recommended resonating the antenna using base or center loading, but specified the resonator must be a part of the antenna, external to the vehicle. The reason for this is readdressed below. Since a resonant antenna's impedance is not usually a match for 50-Ω coax, author Belrose described circuitry to achieve an exact 50-Ω match, using an L-section network. But this additional matching circuitry is an inconvenience for the mobiler who wants to operate on several bands. Besides, most modern transceivers come with a very useful automatic antenna system tuning unit (ASTU), providing the SWR is not too high. Therefore, it is really only necessary to resonate the antenna for minimum SWR with the ASTU switched off, since the automatic ASTU can then correct for the resistance mismatch on the 50-Ω coaxial feed line.

Further, a number of commercially available automatic ASTUs are available. These can be installed close to the antenna feedpoint (say, in the trunk of the automobile). But do not use such a unit to resonate an untuned electrically short mobile whip.[11]

The capacitance C_a of a 110-inch whip at 3.75 MHz is 33 pF. Since the feed-through capacitance C_{bi} of a typical base insulator can be 3-15 pF, an appreciable non-radiating current can flow through it (see **Fig 11A**) if the tuning coil is inside the car. If the antenna is resonated "outside" the metal vehicle, by a base- or center-loading inductor (see Fig 11B), the capacitance of the feedthrough becomes unimportant. Certainly a coaxial cable, no matter how short, should not be used to connect an untuned mobile whip to the tuner. Furthermore, the Q of the inductances used in a compact automatic ASTU will be low with respect to the Q for large diameter air-spaced resonator coils.

The loading coil is an important consideration, since a high-Q is wanted, particularly for 80 meters. The inductance required is very near to the self-resonant frequency for a coil with no space between turns. An air-wound coil with a space between turns should be used, to reduce the self capacity of the coil and to increase its self-resonant frequency. Ideally, a separate coil should be used for each band. An interesting alternative to using a separate coil for each band is described by D. K. Johnson, W6AAQ [reference his application notes for the "Big DK^3 HF Mobile Antenna," or his new book on "HF Mobileering," printed by Worldradio, 1993]. This antenna uses a large lower antenna mast (2-inch diameter), and the series tuning coil moves down inside the lower mast as the frequency is increased—the coil is fully extended at the lowest frequency for the antenna. Integrity is assured by installing fingerstock at the top end of the lower mast section. Although coil turns are in effect eliminated as they go into the pipe mast—there are in effect no shorted turns—they are not turns anymore, since current flows on the top section whip, on the exposed part of the inductor, and on the outside surface of the lower mast section.

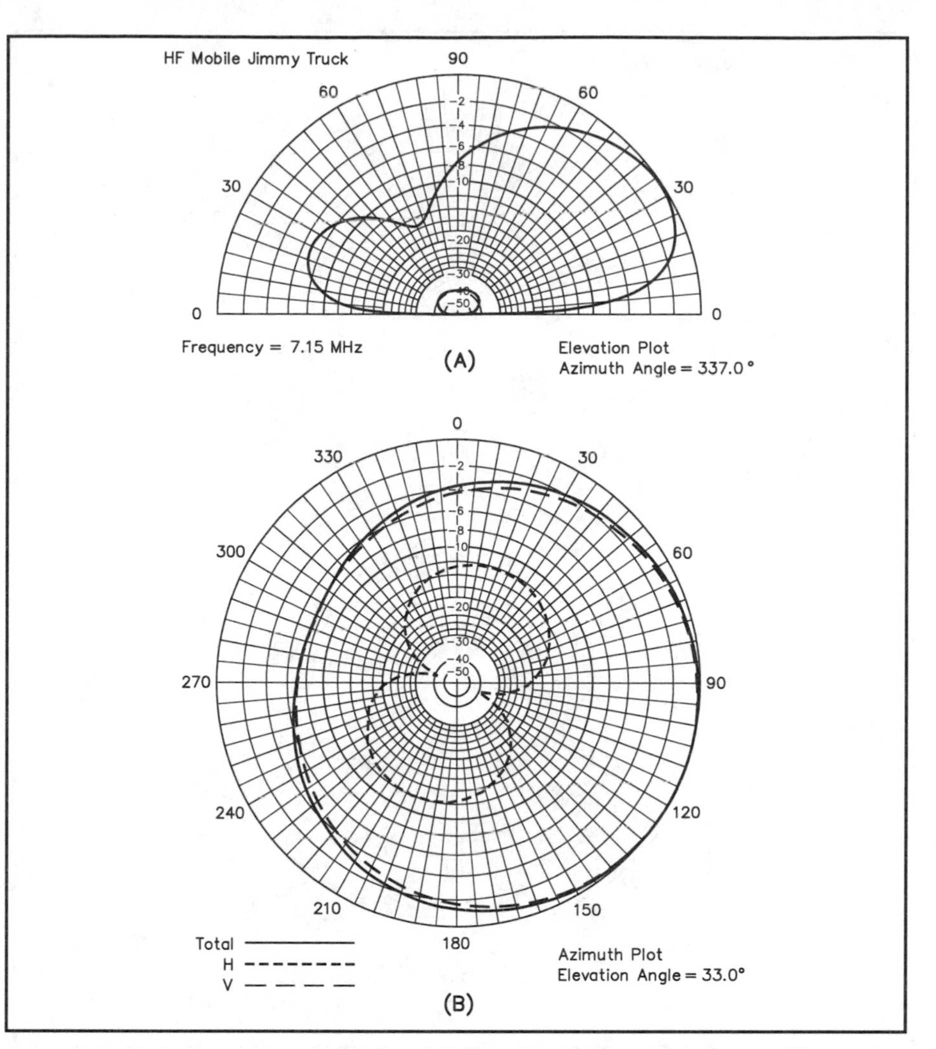

Fig 8—Bumper-mounted whip on GMC Jimmy truck, as for Fig 6, frequency 7.15 MHz.

Concluding Remarks

Mobile antennas are strongly affected by the structure on which they are mounted. This study has shown the patterns to be expected with typical land vehicular mobile antennas. In particular we have emphasized the effect of the close proximity of a bumper-mounted antenna to the back of a 4×4 type vehicle. Clearly, a bumper-mounted mobile antenna on a 4-door sedan with a deep low-profile trunk, or on a pick-up truck, would give better performance, but the difference would only amount to about 2 dB.

The directional azimuth pattern when employing a rear-bumper-mounted HF coil-loaded whip on a rubber-tired vehicle can be used to advantage for communications in a crowded band, but only providing the vehicle is stationary with the desired azimuth orientation, or that it is driven in the direction of propagation! When the vehicle is parked, advantage can be taken of local specifically chosen terrain (for example, a mountaintop or a seaside location with the ocean in front of the antenna). An insulated wire approximately $\lambda/4$ long could be laid on the ground to enhance the directivity in the desired direction.

For NVIS communications using HF antennas on rubber-tired vehicles, an electrically small (compact) loop could be employed.[12] Whatever the mobile platform, rubber-tired or track vehicles, or marine vessels, a compact loop is a better antenna for NVIS.

Acknowledgment

The author works for the Communications Research Centre, Shirleys Bay, ON. He acknowledges the opportunity to use instrumentation and computing facilities not normally available to the average radio amateur.

Notes and References

[1]J. S. Belrose, "Short antennas for mobile operation," *QST*, Sep 1953, pp 30-35, p 108.

[2]R. C. Hansen, "Optimum inductive loading of short whip antennas," *IEEE Trans. Veh. Tech., VT-24*, 1975, p 21.

[3]R. C. Hansen, "Efficiency and matching trade-offs for inductively loaded short antennas," *IEEE Trans. on Comm., COM-23*, April 1975, pp 430-435.

[4]J. S. Belrose, "VLF, LF and MF antennas," in *The Handbook of Antenna Design*, ed. Rudge, Milne, Olver and Knight (London: Peter Peregrinus, 1983), pp 627-630, 633.

[5]The author uses the *ELNEC* version, by Roy Lewallen, PO Box 6685, Beaverton, OR 97007.

[6]The author uses the *NEC/Wires* version, by Brian Beezley, 3532 Linda Vista Dr, San Marcos, CA 92069.

[7]B. F. Brown, "Optimum Design of Short Coil-Loaded High-Frequency Mobile Antennas," *ARRL Antenna Compendium, Vol 1*, 1985, pp 108-115.

[8]M. A. Wallace, "HF Radio in Southwest Asia (during Desert Storm)," *IEEE Communications Magazine*, Oct 1991, pp 58-61.

[9]J. Gold, "HF Mobiling—Taking it to the Streets," *QST*, Dec 1993, pp 67-69.

[10]R. Burch, "You Can Operate HF Mobile!" *QST*, Feb 1993, pp 29-30.

[11]J. S. Belrose, "Automatic Antenna Tuners for Wire Antennas," Technical Correspondence, *QST*, Mar 1994.

[12]J. S. Belrose, "An Up-Date on Compact Loops," *QST*, Nov 1993, pp 37-40.

APPENDIX A

Impedance of a Short Monopole: Simple Analysis

Belrose[1,4] analyzes inductively loaded whips by an equivalent transmission line

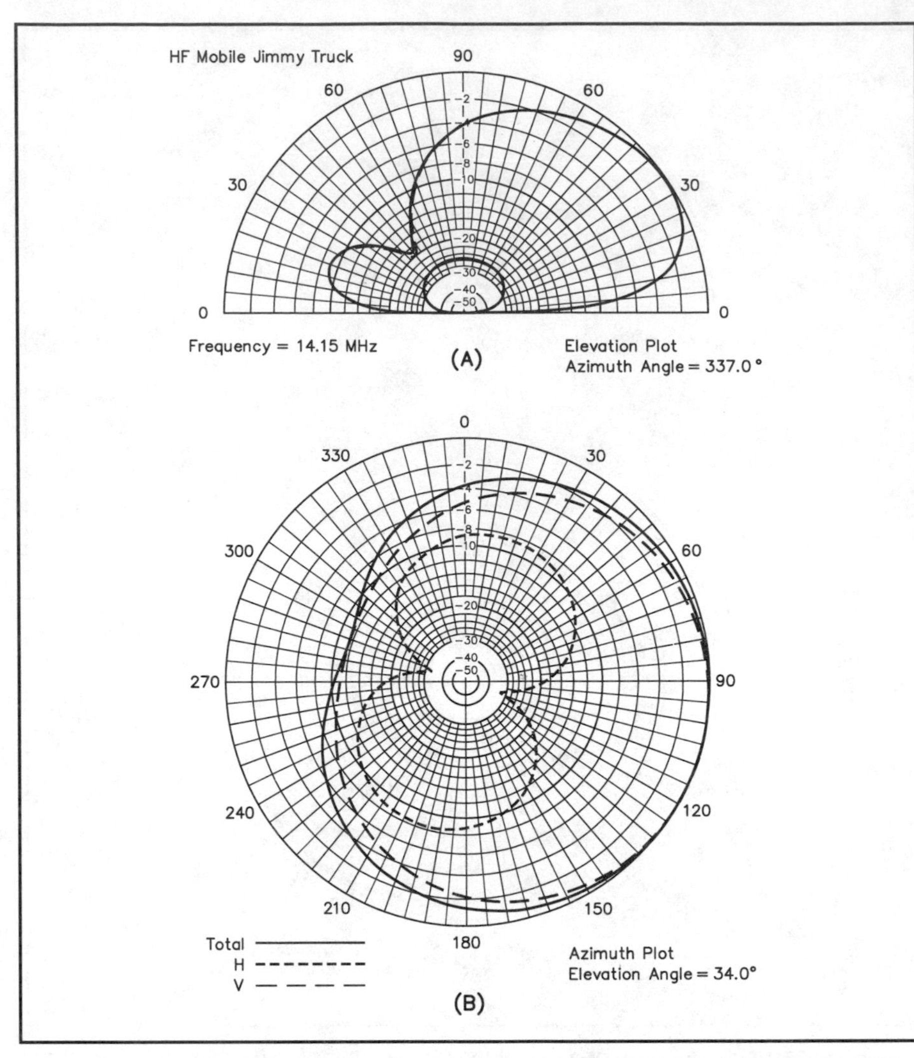

Fig 9—Bumper-mounted whip on GMC Jimmy truck, as for Fig 6, frequency 14.15 MHz.

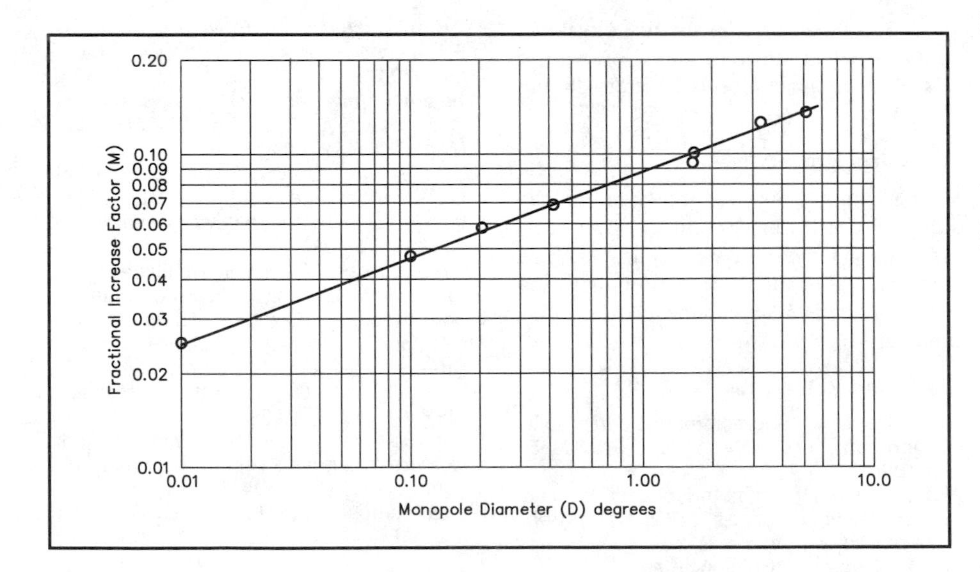

Fig A1—Graph of Fractional Increase Factor (M) versus monopole diameter D in degrees.

method. The earlier method assumed linear current on the antenna. In the later analysis the author assumed a sinusoidal current distribution on the antenna.

The following is a sample calculation for a center-loaded whip for 3.75 MHz. The height of the monopole:

$h = h_1 + h_2 = 2.794$ meters, and the average diameter d = 8 mm.

The inductance L_0 is adjusted so that the base reactance is purely resistive, that is, the antenna is resonant. The lower section of the antenna is considered to be an opened-out transmission line loaded by the capacitive reactance of the top section and the inductive reactance of the loading coil in series. For resonance

$$X_{Lo} = j\,Z_0(\cot G_1 - \cot G_2)$$

where G_1 and G_2 are the electrical heights of h_1 and h_2, and Z_0 is the average characteristic impedance of the monopole of height h.

$$Z_0 = 60\ (\ln\frac{2h}{d} - 1)$$

For $h_1 = h_2 = 1.397$ m

$$G_1 = G_2 = (1 + M)\ \frac{h}{\lambda}\ (360°)$$

$$= \frac{1.035\ (1.397)\ (360°)}{80} = 6.5°$$

The factor $(1 + M) = 1.035$ is the reciprocal of the usual antenna factor k. Information on the relation between M and the electrical diameter in degrees (D) of the whip is shown in **Fig A1**. For our mobile whip, the electrical diameter D:

$$D = \frac{d}{\lambda}\ (360°) + \frac{8(360°)}{80 \times 10^3} = 0.036°$$

Hence M = 0.035. The characteristic impedance

$$Z_0 = 60\ (\ln\frac{2(2.794)}{8} - 1) = 333\ \Omega$$

and $X_{Lo} = j333\,(\cot 6.5° - \tan 6.5°) = j2922\,\Omega$

The radiation resistance can be calculated by adding the current areas (degree-amperes) on the two parts of the antenna, since the currents are in phase.

$$A_1 = \frac{180}{\pi}\left[\frac{1 - \cos G_1}{\sin G_2}\right]\cos G_1$$

= 3.23 degree-amperes, and

$$A_2 = \frac{180}{\pi}\sin G_2$$

= 6.49

Hence $R_r = 0.01215\,(A_1 + A_2)^2 = 1.15\,\Omega$. These values can be compared with those calculated by *ELNEC* in Table 1.

Fig 10—A long mobile whip (20 feet) tilted over the top of the vehicle: at A, wire model and currents for a frequency of 3.75 MHz; at B, elevation pattern for 0° azimuth; and at C, principal plane azimuth pattern.

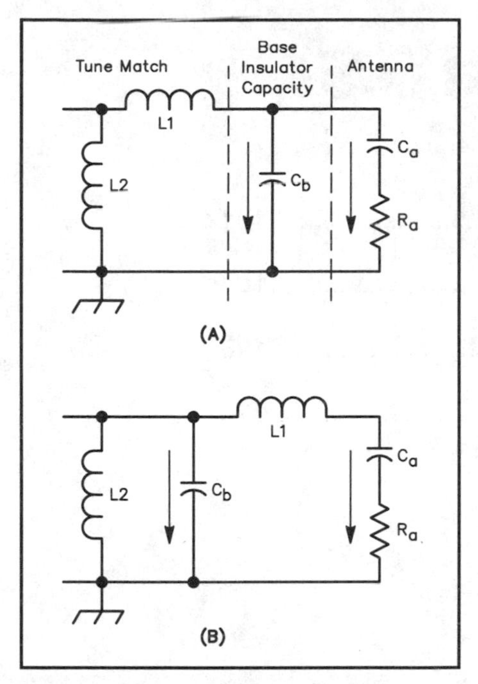

Fig 11—Equivalent circuit illustrating incorrect (at A) and correct method (at B) of tuning an electrically short monopole, where a base insulator feedthrough to the inside of a metal vehicle or an ASTU is employed.

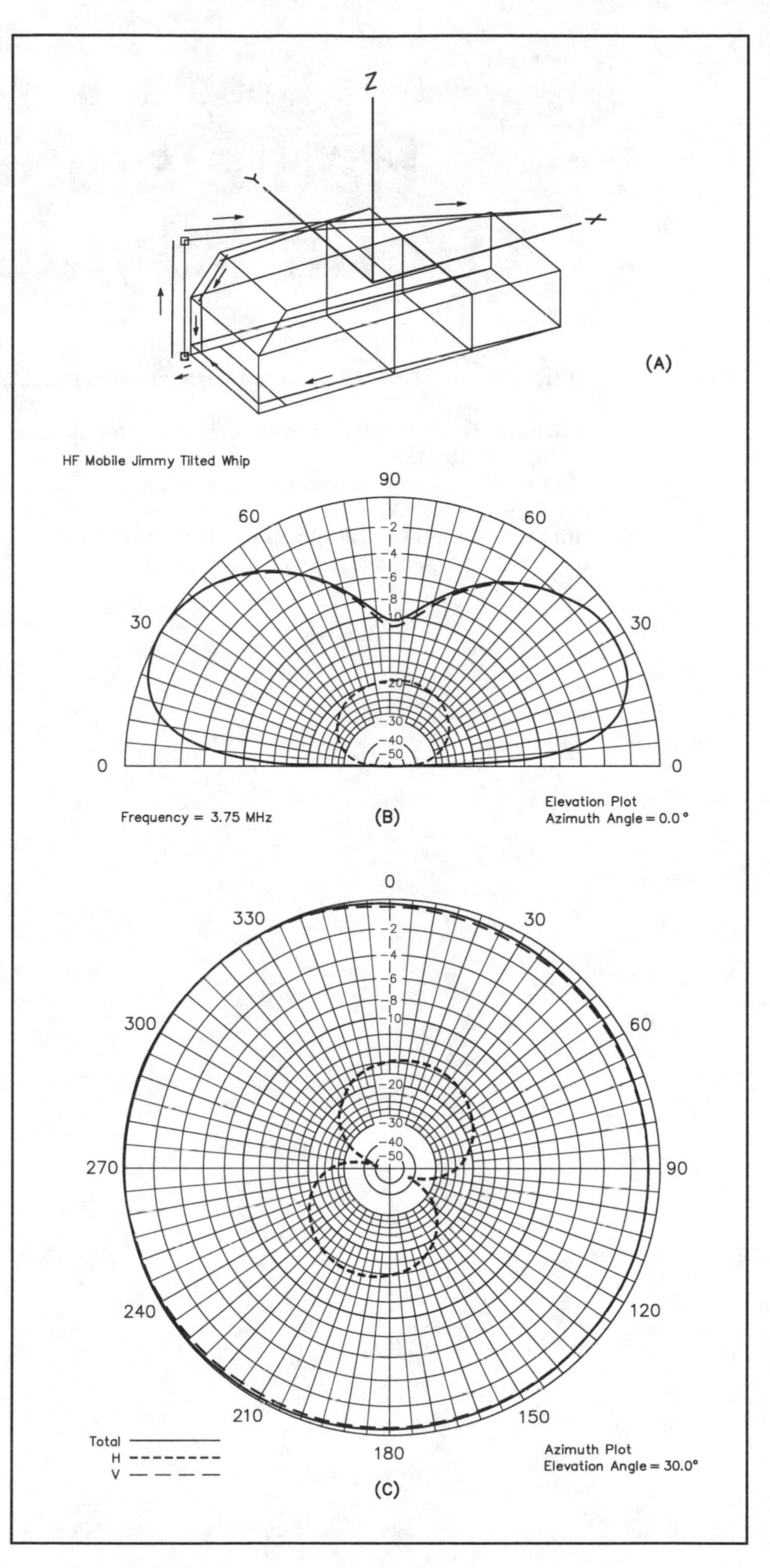

Chapter 6
Radials and Ground Systems

from June 1985 *QST*

Radial Systems for Ground-Mounted Vertical Antennas

Thinking about putting up a vertical antenna? This modern Numerical Electromagnetic Code-Method of Moments computer study will give you an idea how many radials to put under it, and how long they should be.

By Brian Edward,* N2MF

Ground-mounted vertical antennas are a popular choice for amateur communications, particularly on the lower HF bands. There are many reasons for their popularity. These antennas are simple to erect, easy to tune, are relatively unobtrusive and produce low-angle radiation. Vertical antennas seem to appeal especially to two groups of amateurs: the newcomer and the low-band DXer. The newcomer appreciates the first three traits of these antennas, while the DXer recognizes the potential of the antenna for providing competitive performance on long-haul contacts, particularly when employed as an element in an array.[1]

For the ground-mounted vertical antenna to perform properly, it must be used in conjunction with a counterpoise in which the antenna image currents flow. This counterpoise usually consists of an arrangement of radial wires in combination with the surface of the earth. The question arises as to how many radial wires are needed and what length they should be. A common answer is, "The more the better, and make them as long as you can." A little research into this topic, however, reveals that there is a trade-off between the number of radials and the radial length. If only a few radials are going to be used, they need not be very long. If more radials are added, the length of all radials should be extended in order to obtain the full performance potential of the antenna system.

A Study of Radial Systems

John O. Stanley has written an excellent article giving design information for radial ground systems.[2] He suggested radial lengths to be used for a given number of radial wires. An extensive experimental study on this topic was also performed by Brown and others. Their work was documented as early as 1937.[3] I studied this subject using the Numerical Electromagnetic Code (NEC)-Method of Moments computer program.[4] NEC is a powerful program that allows you to analyze wire antennas in the presence of actual ground conditions. Through this study, I was able to determine the performance of various radial systems in combination with different earth electrical characteristics. The program calculates relative gain for each system studied, along with the elevation angle (measured from the horizon) at which maximum radiation occurs.

My study was performed with radial wire numbers (N) of 4, 12, 24, 48, 96 and 120. The radial lengths for each system ranged from 0.05 to 0.6 wavelength (λ). As a first case, I selected a system with four radial wires, considered to be what the typical amateur might choose as a starting point for a vertical-antenna ground system. The case with 120 radials (at lengths of 0.5 λ) is considered the optimum arrangement for medium-frequency broadcast-station antennas.

Each radial system was studied in combination with earth characteristics that may be considered poor, good and very good. These characteristics are determined by the soil conductivity σ, in siemens/meter, and the relative dielectric constant, ϵ_r. Typical values for these parameters are given in Table 1. It turns out that the ratio of conductivity to frequency is the first parameter of importance. For the results reported in this article, the ratio is given by $X = \sigma/f$, where f is the frequency in megahertz. The three earth characteristics were then nominally chosen to correspond to the values shown in Table 2.

By comparison to Table 1, you may note that the conditions that are often called poor earth characteristics are not all that far removed from the average soil characteristics where many of us live. I should also point out that earth characteristics often vary widely over distances of a few feet and over a period of several months.[5] It could be interesting to measure the actual soil conductivity in the vicinity of your antenna site at different times of the year.

My study was performed for the sky-wave component of the antenna radiation

Table 1
Typical Earth Electrical Characteristics†

Terrain	*Conductivity (Siemens/Meter)*	*Relative Dielectric Constant (ϵ_r)*
Seawater	5	80
Fresh Water	0.008	80
Dry, sandy, flat coastal land	0.002	10
Marshy, forested flat land	0.008	12
Rich agricultural land, low hills	0.01	15
Pastoral land, medium hills and forestation	0.005	13
Rocky land, steep hills	0.002	10
Mountainous	0.001	5
Cities, residential areas	0.002	5
Cities, industrial areas	0.001	3

†Information adapted from *Reference Data for Radio Engineers* (Indianapolis: Howard W. Sams & Co., 1979, p. 28-3.

*3000 Henneberry Rd Jamesville, NY 13078

Table 2
Ground Characteristics Used for Study

Ground	$X \left(\frac{\text{Siemens/Meter}}{\text{MHz}} \right)$	ϵ_r
Poor	0.0001	7
Good	0.001	15
Very good	0.01	30

as opposed to the ground- or surface-wave component. The sky wave is the component of interest to amateurs communicating via the ionosphere, while medium-frequency broadcasters are more interested in the surface-wave component. I used a quarter-wavelength-long resonant structure as the vertical radiating element. The radials were no. 12 wire, although the actual wire size is of little importance (see note 3).

Results

The results of my study are presented in graphical form. Figs. 1, 2 and 3 show the gain of the antenna systems (the gain of a half-wavelength dipole in free space is 2.15 dB, with respect to an isotropic radiator) for the various radial configurations operating in combination with poor, good and very good earth characteristics. Don't become too concerned with the actual gain numbers. Instead, examine the relative gains provided by the different ground systems. You can see that for all cases, if relatively short radials are to be used, there is no need to use many of them. For example, with poor earth, if the radial length must be restricted to 0.1 λ, then 24 wires is the maximum that need to be used. Alternatively, if a large number of radials is to be used, they should be long in order to realize the maximum antenna-system performance. These results agree with those presented by Stanley and by Brown, Lewis and Epstein. Although the graphs present results for specific numbers of radials and specific earth characteristics, you should be able to interpolate for other radial numbers and earth characteristics.

You can also see that for very good earth electrical characteristics (Fig. 3), it doesn't take very many radial wires to obtain good performance. This has been proven by the big signals emanating from the Caribbean by suitcase DX peditioners operating on the beach with simple vertical antennas.

Figs. 4, 5 and 6 show the elevation angles for maximum radiation with the various radial configurations and earth characteristics. This elevation angle is determined largely by the earth characteristics a wavelength or more beyond the vertical radiator—in other words, beyond the typical radial system. This shows up in the plots where the lowest elevation angles correspond to the best earth characteristics. The elevation angle can be lowered somewhat when poorer earth characteristics are present by employing an extensive radial system. If a perfectly conducting, infinite-size ground plane were available, the elevation angle of maximum radiation would be zero degrees (at the horizon).

The graphs of Figs. 1 through 3, corresponding to the three earth characteristics studied, were used to determine a sufficient radial length for a given number of radial wires. Two similar criteria were used to determine the optimum lengths. The first specifies a radial length for a given number of wires when the system gain is within 0.1 dB of the maximum gain possible with that number of wires and earth characteristics. The second criterion specifies a radial length for which the gain is within 0.2 dB of the maximum value for that number of wires and earth characteristics. When these radial lengths were tabulated and compared for the three earth characteristics studied, I found that the lengths for a given number of radial wires are not strongly dependent on the earth characteristics. This, of course, does not imply that a given radial system performs the same when used in combination with different earth characteristics. A poorer earth must be compensated for with

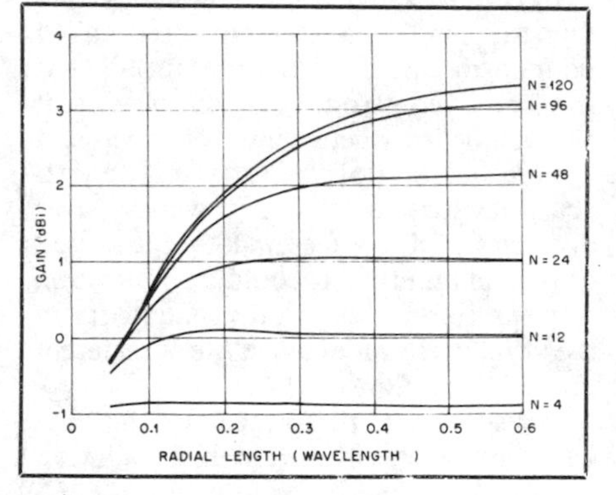

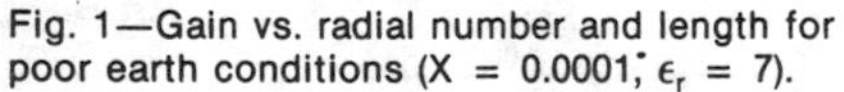
Fig. 1—Gain vs. radial number and length for poor earth conditions (X = 0.0001; ϵ_r = 7).

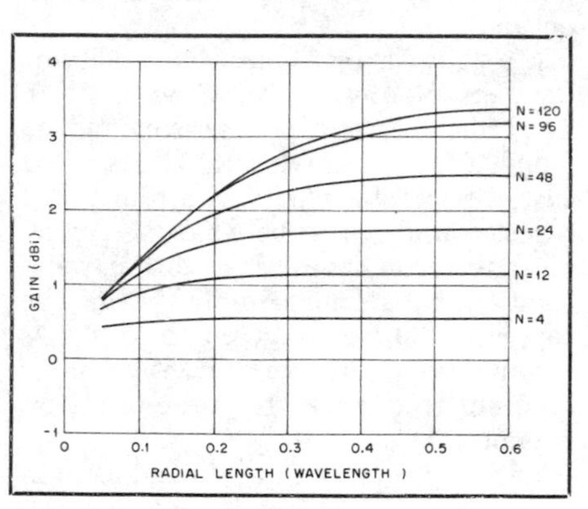

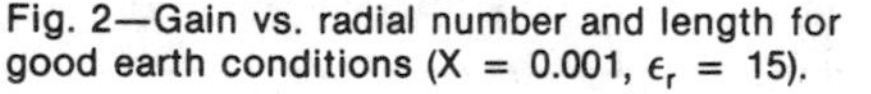
Fig. 2—Gain vs. radial number and length for good earth conditions (X = 0.001, ϵ_r = 15).

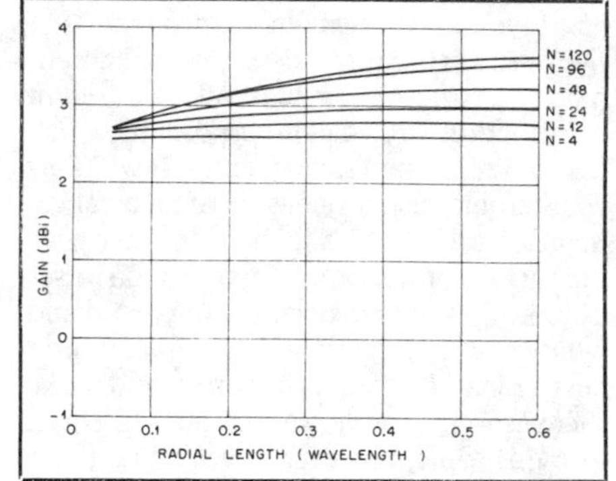

Fig. 3—Gain vs. radial number and length for very good earth conditions (X = 0.01, ϵ_r = 30).

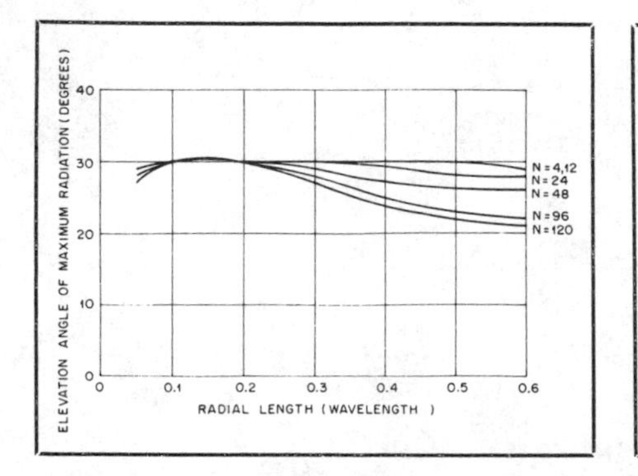

Fig. 4—Elevation angle vs. radial number and length for poor earth conditions (X = 0.0001, ϵ_r = 7).

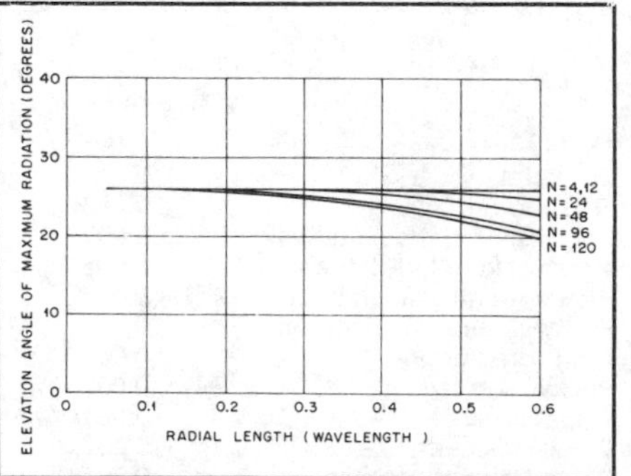

Fig. 5—Elevation angle vs. radial number and length for good earth conditions (X = 0.001, ϵ_r = 15).

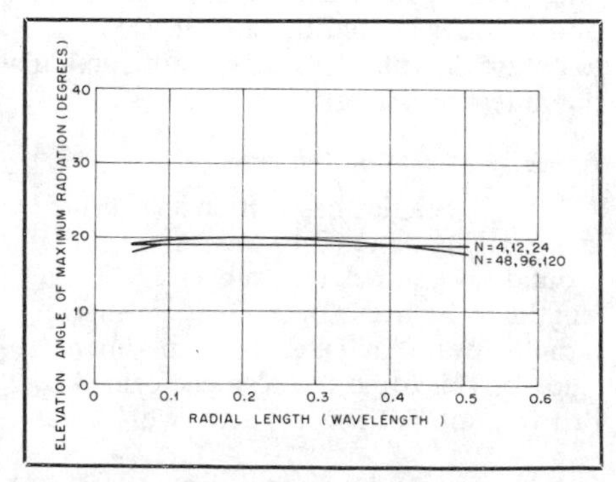

Fig. 6—Elevation angle vs. radial number and length for very good earth conditions (X = 0.01, ϵ_r = 30).

more and longer radials to equal the performance of a less extensive system over good earth.

A sufficient wire length for a given number of radials is plotted in Fig. 7. The two upper curves correspond to the criteria explained above. The lower curve is the radial wire lengths given by Stanley. (See note 2.) His radial lengths are somewhat shorter than the ones I found. Possibly, the criteria he used for choosing the lengths were not as well defined as mine. It might also be possible that his data was for a surface-wave study, which would be of more interest to medium-frequency broadcasters than would a sky-wave study. You may also note that there is an apparant discontinuity between his second-to-last data point, which is for 90 radial wires, and his last point, which is for 120 wires.

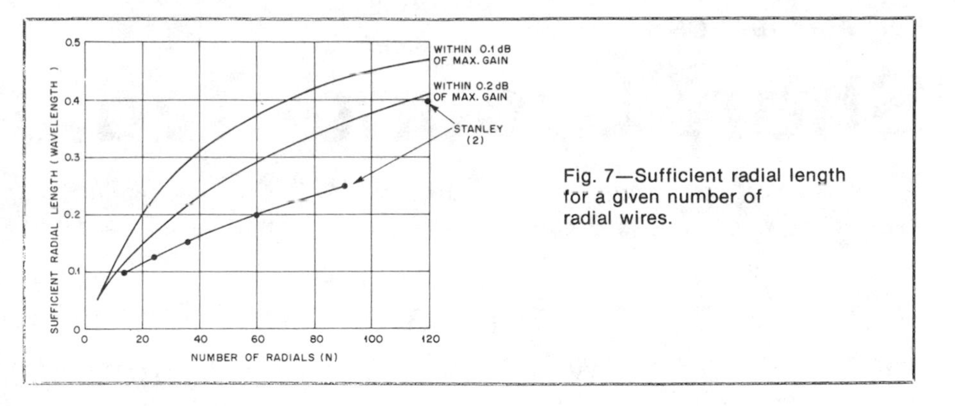

Fig. 7—Sufficient radial length for a given number of radial wires.

As stated earlier, the ground systems studied were operating in conjunction with a quarter-wavelength-long vertical radiator. The gains given by Figs. 1 through 3 will essentially hold for vertical-antenna lengths down to approximately an eighth of a wavelength. For even shorter antennas, the gains will tend to decrease. Therefore, with short verticals, more extensive ground systems should be employed to maximize the system performance.

Conclusions

The performance of a vertical antenna operating in conjunction with a radial ground system has been studied. The effect of a number of radial wires, wire length, and the electrical characteristics of the earth on the relative gain of an antenna system and the elevation angle at which the maximum gain occurs has been determined. For a given number of radial wires, there is a corresponding sufficient wire length, which is, surprisingly, independent of the earth characteristics. Extending the radials beyond this sufficient length without adding additional radials will yield no substantial performance improvement.

This study has also shown that the elevation angle of maximum gain is largely determined by the electrical characteristics of the earth surrounding the antenna system. This elevation angle can be lowered somewhat by employing more extensive radial systems.

For a given number of radial wires, the sufficient length may be determined by using the graph in Fig. 7. Then, by measuring or estimating (from Table 1) the earth's electrical characteristics and calculating the parameter X for the operating frequency, you can determine the performance of the vertical antenna with the radial ground system using Figs. 1 through 6.

Notes

[1]J. C. Rautio, "The Effect of Real Ground on Antennas," Parts 1-5, *QST,* Feb., April, June, Aug. and Nov. 1984.

[2]J. O. Stanley, "Optimum Ground Systems for Vertical Antennas," *QST,* Dec. 1976, pp. 13-15.

[3]Brown, Lewis and Epstein, "Ground Systems as a Factor in Antenna Efficiency," *Proc. of the IRE,* June 1937.

[4]Numerical Electromagnetics Code-Method of Moments, Developed by Burke and Poggio, Lawrence Livermore Laboratory.

[5]A. C. Doty, J. A. Frey and H. J. Mills, "Efficient Ground Systems for Vertical Antennas," *QST,* Feb. 1983, pp. 20-25.

from April 1978 *QST*

Short Ground-Radial Systems for Short Verticals

When is a ground not a ground? Should my radials be buried? How deep? How many? Will my vertical work without a ground? Let W2FMI give you the answers.

By Jerry Sevick, W2FMI

How do you engineer the performance of ground-radial systems under vertical antennas? There isn't much engineering design information available, particularly for conditions where space is limited and cost is an important consideration.

The often-asked questions which need answering are (a) Do four quarter-wavelength radials constitute an adequate ground system? (b) Must radials be buried deeply in the earth? (c) Must the thickest copper conductor available be used? And while we are at it, how about the mistaken notion that short verticals can never compete in performance with a full-sized quarter-wavelength antenna?

This paper presents experimental evidence which answers these questions in a clear and concise way. Investigation shows that short verticals over very small radial systems of almost any kind of thin wire on the ground's surface can perform surprisingly well. On-the-air comparisons with much larger verticals over extensive ground systems show performance reduced by only a few decibels.

Introduction

Vertical antennas have enjoyed considerable popularity on the 80- and 160-meter amateur bands because of the difficulty of erecting horizontal antennas at heights sufficient for low-angle radiation. Optimum heights, which are in excess of a half wavelength (λ) on these bands, are impractical for most amateurs. In many cases short verticals are used since they have been shown to compete favorably with 1/4-λ vertical antennas.[1] This is true if losses in the ground system, matching networks and loading elements are small compared to the reduced radiation resistances of the short antennas.[2] Considerable information is available describing the effects of buried radials on the efficiency of 1/4-λ verticals in the mf and lf bands as a function of length and number of radials, and conductivity of the soil.[3-19] But little is available on radials lying on the ground's surface, particularly in connection with short verticals.

During the author's experiments, various antenna heights from 1/4 λ to 1/8 λ were included. Also developed and described here is a simple method for measuring an important parameter for vertical antennas — soil conductivity. The conductivity of the soil under and in the near vicinity of the antenna is most important in determining the extent of the radial system required and the overall performance. As will be seen, short verticals with very small radial systems can be surprisingly effective.

Soil Conductivity

Most soils are nonconductors of electricity when completely dry. Conduction through the soil results from conduction through the water held in the soil. Thus, conduction is electrolytic. Dc techniques for measuring conductivity are impractical because they tend to deplete the carriers of electricity in the vicinity of the electrodes. The main factors contributing to the conductivity of soil are

1) Type of soil.
2) Type of salts contained in the water.
3) Concentration of salts dissolved in the contained water.
4) Moisture content.

Elements of the 10 vertical antennas used on 20 and 40 meters to experimentally determine the efficiency of shortened verticals with abbreviated radial systems.

5) Grain size and distribution of material.

6) Temperature.

7) Packing density and pressure.

Although the type of soil is an important factor in determining its conductivity, rather large variations can take place between locations because of the other factors involved. Generally, loams and garden soils have the highest conductivities. These are followed in order by clays, sand and gravel. Soils have been classified according to conductivity, as shown in Table 1. Although some differences are noted in the reporting [20,21] of this mode of classification because of the many variables involved, the classification generally follows the values shown in the table.

Table 1
General Classification in Conductivity

Material	*Conductivity (millimhos/meter)*
Poor Soil	1-5
Average Soil	10-15
Very Good Soil	100
Salt Water	5000
Fresh Water	10-15

Since conduction through the soil is almost entirely electrolytic, ac measurement techniques are preferred. Many commercial instruments employing ac techniques are available and described in the literature.[22] But rather simple ac measurement techniques can be used which provide accuracies on the order of 25 percent and are quite adequate for the radio amateur. Such a setup was developed by a colleague and neighbor, M. C. Waltz,[23] W2FNQ and is shown schematically in Fig. 1. Fig. 2 shows the conductivity readings taken over the last three months in 1976. It is interesting to note the general drop in conductivity over the three months as well as the short-term changes due to periods of rain. The results presented in the following sections on antenna efficiencies were obtained in the period October 10 to November 10, 1976, when the conductivity varied between 22 and 25 millimhos/meter.

Antenna Efficiency Considerations

The antenna efficiencies to be discussed are based upon the losses which appear in series with the radiation resistance of resonant verticals. Although this approach does not give a comparison between the very low angles of radiation (i.e., less than 15 degrees) of various radial systems, it does allow for comparisons in the 15- to 30-degree range which is important for sky-wave transmission on the 40-, 80- and 160-meter bands. Mathematically this definition for antenna efficiency can be written as

$$\text{Antenna efficiency} = \frac{R_{rad}}{R_{rad} + R_g + R_A}$$

where R_{rad} = radiation resistance

R_g = ground loss

R_A = ohmic losses due to loading and the antenna itself.

With high-Q loading coils and practically any size of aluminum tubing for the antenna, R_A can be minimized and therefore eliminated from the relationship above.

An example of this technique for determining antenna efficiency uses the results shown in Fig. 3. The input impedance of a resonant quarter-wavelength vertical is plotted as a function of the number of radials. Two lengths of radials (0.2 λ and 0.4 λ) were considered. Since the radiation resistance is 35 ohms for the thickness of the verticals used in this experiment, it can be seen that with 50 radials, losses were about 2 ohms and with 100 radials, 1 ohm. This amounts to efficiencies of 94 and 97 percent, respectively. Also, it can be seen that the efficiency with only four radials is less than 60 percent. This poor efficiency exists even for a location with a soil conductivity that can be considered average.

Further, the efficiency of a radial system employing small numbers of radials is quite dependent on the moisture content of the soil. Fig. 4 shows this result with a resonant quarter-wavelength vertical on 20 meters while the number of radials varies from one to eight. As can be seen, the difference in efficiency between wet and dry conditions becomes less pronounced as the number of radials is increased. The antenna system also becomes more independent of soil conductivity as the number of radials is increased.

In order to determine the efficiencies of shortened verticals over abbreviated radial systems, input impedances were compared with similar antennas over a near-ideal radial system. Fig. 5 shows the experimental results[24] obtained by the author on a near-ideal image plane (100 radials on the ground, about 50 feet long, and terminated in 10- to 12-inch nails). The top-hat loading consisted of an eight-spoked wheel with several rings of aluminum wire to improve its effect. This family of curves has been very useful to the author in designing verticals since it predicts the value of the input impedance of shortened verticals using various loading methods. In particular, it was noted that a simple rule of thumb existed for top-hat loading. That is, a top hat with a diameter D is equivalent to an electrical height of 2D. Further, top-hat loading yielded the highest impedance and bandwidth for a particular height.

Since the investigations reported here involved short radials on the surface of the ground, a study of the effect of the length of a spike terminating the radials was necessary. The efficiency for resonant 1/4-λ verticals with small numbers of

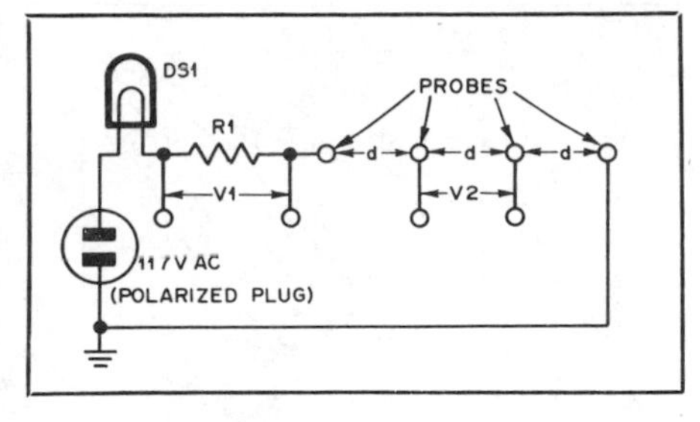

Fig. 1 — Schematic diagram of four-point probe method for measuring earth conductivity.

DS1 — 100-watt light bulb.

R1 — 14.6 ohms (5 watt).

Probes — 5/8-inch dia (iron or copper); spacing, d = 18 inches; penetration depth, D = 12 inches.

$$\text{Earth conductivity} = (21) \times \frac{V1}{V2}$$

(millimhos/meter).

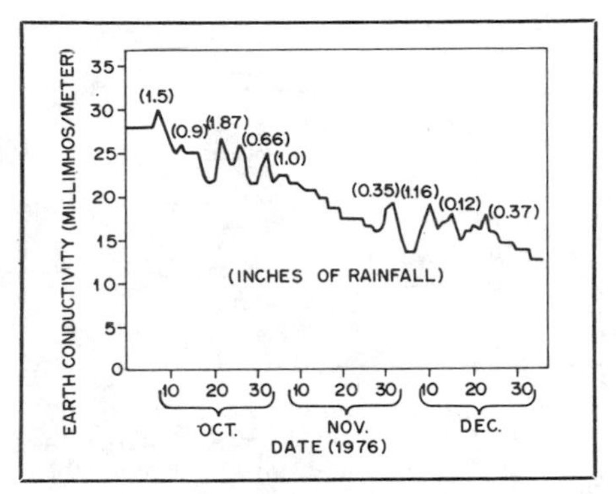

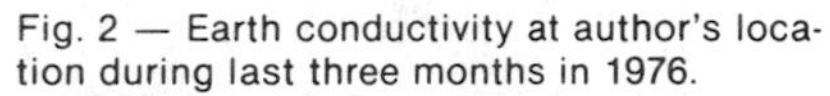

Fig. 2 — Earth conductivity at author's location during last three months in 1976.

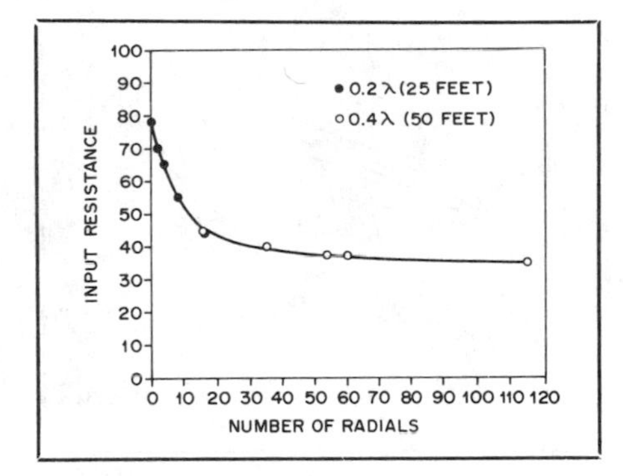

Fig. 3 — Input impedance of resonant quarter-wavelength vertical as a function of the number of radials.

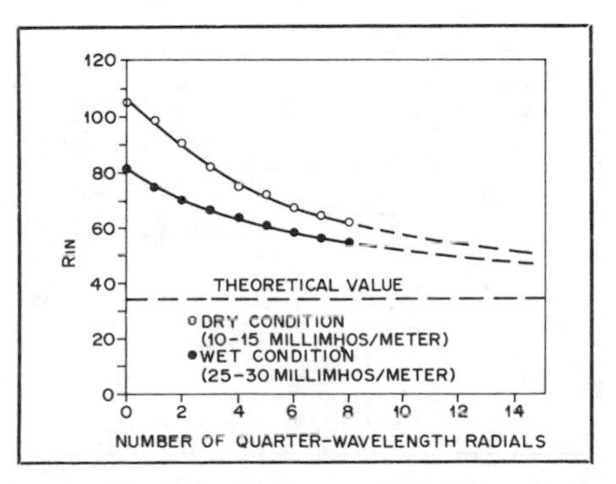

Fig. 4 — Input impedance of resonant quarter-wavelength vertical as a function of the number of radials and the condition of the soil.

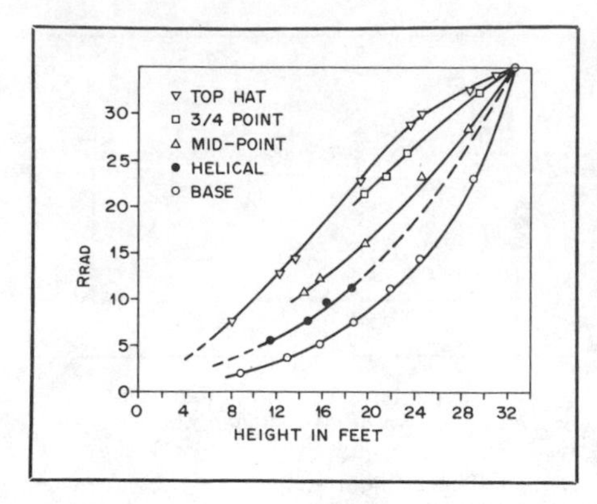

Fig. 5 — Experimental results of radiation resistance as a function of height of antenna for various methods of loading.

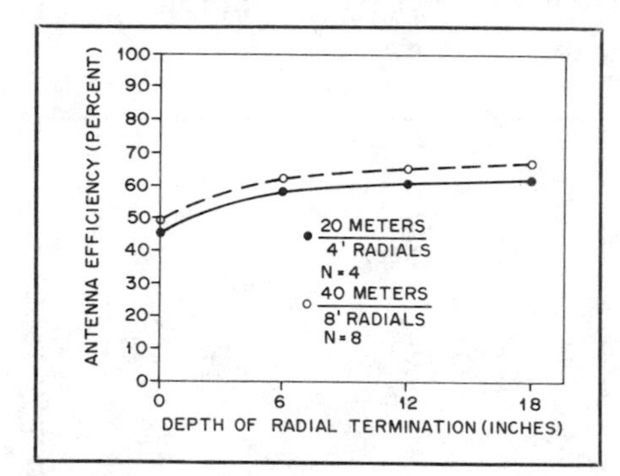

Fig. 6 — Efficiency of resonant quarter-wavelength verticals on 20 and 40 meters as a function of the length of spike terminating the radials.

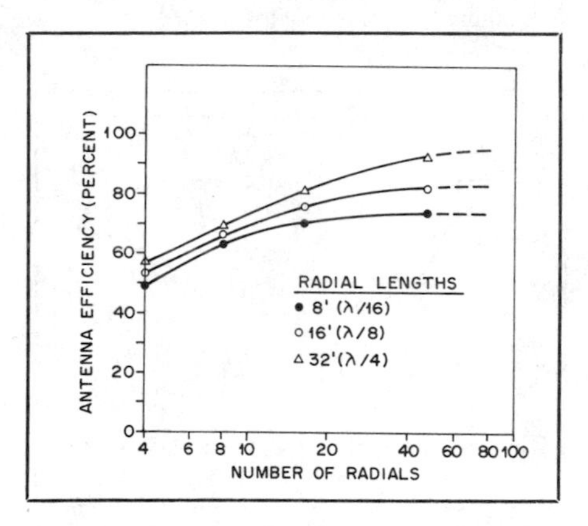

Fig. 7 — Efficiency of resonant quarter-wavelength vertical as a function of the number of terminated radials with three different lengths.

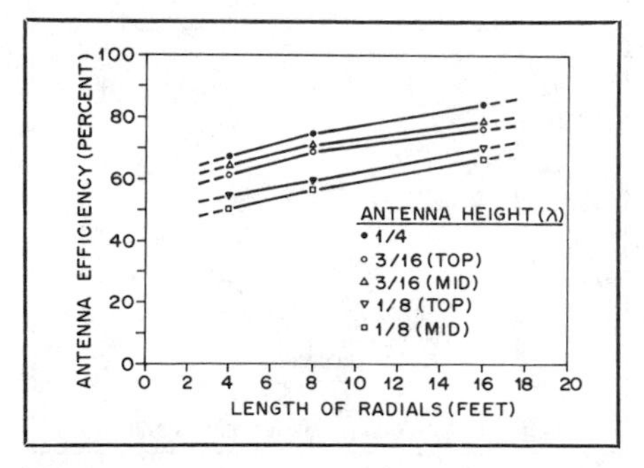

Fig. 8 — Efficiency of short verticals as a function of the length of the terminated radials, with the number kept constant at 48.

shortened radials was measured on the 20-and 40-meter bands. The radials were four and eight feet long (1/16 λ) respectively. Fig. 6 shows the results for four different depths of termination. As can be seen, depths of 10 to 12 inches should be sufficient for the soil conductivity at the author's location for 20 and 40 meters, and most likely for 80 and 160 meters as well. Incidentally, the effectiveness of radials of a 1/4 λ or longer did not change appreciably as a function of the depth of the termination. Therefore, terminations for long radials are primarily used for mechanical reasons.

Abbreviated Radial Systems

In order to determine experimentally the efficiency of shortened verticals with abbreviated radials on the surface of the ground, five verticals of different heights and loading schemes were used on the 20-and 40-meter bands. The results were then compared with similar antennas on a near-ideal image plane as shown in Fig. 5. The five resonant verticals selected were

1) 1/4 wavelength.
2) 3/16 wavelength, top-hat loaded.
3) 3/16 wavelength, midpoint loaded.
4) 1/8 wavelength, top-hat loaded.
5) 1/8 wavelength, midpoint loaded.

One of the photographs shows the elements for these 10 antennas.

A radial system with various lengths of 17-gauge steel electric fence wire, terminated with 10- to 12-inch nails, was employed. A picture shows the 12-inch aluminum base and the input connection arrangement. The antenna system was erected in the front yard about 50 feet from the house, and it offered an opportunity for on-the-air comparisons with verticals mounted on the near-ideal system in the backyard.

The results are shown in Figs. 7 and 8. Only the 40-meter data is presented since little difference was noted on 20 meters. Fig. 7 shows the effect on the efficiency of a resonant 1/4-λ vertical on 40 meters as a function of the number of radials using three different lengths of terminated radials. Although these curves were taken on a 40-meter system, the relationships are generally valid for all other frequencies in the hf range if the same fractional wavelengths are used for the radials. As can be seen the longer radials (1/4 λ) yielded the highest efficiency. But it is interesting to note that this improvement in efficiency with length decreases as the number of radials becomes smaller. At four radials, the efficiency with eight-foot radials is not much poorer than that with 32-foot radials, i.e., 50 percent compared to 56. Further, other interesting trade-offs exist with various lengths and numbers of radials. Fig. 8 shows that 16 1/4-λ radials are equivalent to about 35 1/8-λ radials, and 8 1/4-λ radials are equivalent to only 12 1/16-λ radials. Obviously other equivalences can be obtained from the figure.

Fig. 8 shows the results of antenna efficiency for short verticals as a function of the length of radials, with the number of radials kept constant at 48. As expected, 1/4-λ verticals, with their higher radiation resistance, have the highest efficiencies. Surprisingly, the efficiency of the 1/8-λ verticals does not suffer proportionally. That is, the 1/8-λ vertical with midpoint loading still has a 67-percent efficiency compared to the 84-percent efficiency of the 1/4-λ vertical, even though its radiation resistance is only about one-third as large (12.5 ohms compared to 35 ohms). A further comparison with a 1/4-λ vertical over an ideal image plane predicts that this 1/8-λ vertical with 48 1/8-λ radials (terminated) should show a reduced performance of only 1.7 dB.

On-the-Air Comparisons

As was shown in the previous section, short verticals with a sufficient number of abbreviated radials that are terminated should yield performances only a decibel or two poorer than 1/4-λ verticals over extensive ground systems. Several on-the-air comparisons were made to confirm this prediction, which was based upon efficiency considerations.

The first comparison involved a 40-meter, 1/8-λ, top-hat-loaded vertical with 48 1/8-λ radials (17-gauge steel wire on the ground and terminated with 10- to 12-inch nails). The input impedance of this vertical was about 25 ohms and it was matched to the 50-ohm transmission line with a highly efficient 2:1 step-down transmission-line transformer. This antenna system was compared with a 29-foot vertical using a 13-1/2-foot-diameter top hat in the backyard. The ground system for this larger antenna consisted of 100 radials of no. 15 aluminum wire, each about 50 feet long. Each radial was on the surface of the ground and terminated with 10- to 12-inch nails. This larger antenna was resonated by a small variable capacitor in series at the base. Over 100 contacts on 40 meters plus about 200 observations on reception showed that the differences between the two systems were generally negligible. A few reports showed a 1- to 2-dB difference in favor of the larger system but these were in the minority.

An even more interesting comparison was made on 80 meters. A 20-foot vertical with an 8-foot top hat was erected over the same image plane of 48 16-foot radials in the front yard. It required a base-loading coil of about 20 turns of 12-gauge wire, 2-1/2-inch diameter, 6 tpi to resonate it. The input impedance was about 12 ohms and a 4:1 transmission line transformer was used for matching. This antenna configuration represents a radiation resistance of about 5 ohms and a loss in the ground system and loading coil of about 7 ohms. The 29-foot vertical with the 13-1/2-foot diameter top hat needed

The 12-inch aluminum base and input-connection arrangement. Shown are 48 radials of 17-gauge steel electric-fence wire.

about eight turns on a powder-iron core (T200) at the base to resonate it on 80 meters. Its input impedance was 15 ohms (showing negligible loss in the ground system and base loading coil), and matching was accomplished with an efficient transmission-line transformer having a 3.33:1 step-down impedance ratio. Again about 100 contacts were made on the air and another 200 observations were made on reception. The average difference between these two systems amounted to only about 5 dB in favor of the much larger system in the backyard. This is quite noteworthy since many previous contacts with the larger antenna established it as a very competitive antenna system.

Concluding Remarks

Quarter-wavelength vertical antennas over an extensive radial ground system have been known to be efficient, low-angle radiators. Even short verticals over the same large ground system have been shown to lose little in the way of performance. With low-loss matching and loading techniques, short verticals over a large ground system suffer only in bandwidth. But full-sized verticals and radial ground systems are beyond the reach of most radio amateurs on the 80- and 160-meter bands. This investigation was undertaken because little information was available on limited radial systems, particularly for short verticals.

As was shown, short radials over soil of average conductivity can perform quite acceptably for verticals of all heights. The results of this investigation now allow one to predict quite accurately the operation of verticals with heights less than a quarter-wavelength and with radials as short as 1/16 λ in length. The simple soil-conductivity measurement scheme described also gives one a tool for comparing a given location with others, as well as predicting the performance of a vertical antenna system.

References

[1] Sevick, "The W2FMI Ground-Mounted Short Vertical," *QST*, March, 1973.
[2] Sevick, "Broadband Matching Transformers Can Handle Many Kilowatts," *Electronics*, November 25, 1976.
[3] Brown, Lewis and Epstein, "Ground Systems as a Factor in Antenna Efficiency," *Proc. IRE 25*, pp. 753-787, June, 1937.
[4] Abbott, "Design of Optimum Buried R.F. Ground Systems," *Proc. IRE 40*, pp. 846-852, July, 1952.
[5] Wait and Pope, "Input Resistance of LF Unipole Aerials," *Wireless Engineer 32*, pp. 131-138, May, 1955.
[6] Wait and Pope, "The Characteristics of a Vertical Antenna With a Radial Conductor Ground System," *Appl. Sci. Research (B) 4*, pp. 177-195, 1954.
[7] Monteath, "The Effect of the Ground Constants and of an Earth System on the Performance of a Vertical Medium-Wave Aerial," *Proc. IEE*, Vol. 105C, Pt. 2, pp. 292-306, January, 1958,
[8] Smith and Devaney, "Fields in Electrically Short Ground Systems: An Experimental Study," *J. Res. NBS* 63D Radio Prop. No. 2, pp. 175-180, September-October, 1959.
[9] Larsen, "The E-Field and H-Field Losses Around Antennas With a Radial Ground Wire System," *J. Res. NBS* 66D Radio Prop. No. 2, pp. 189-204, March-April, 1962.
[10] Maley and King, "Impedance of a Monopole Antenna With a Circular Conducting-Disk Ground System on the Surface of a Lossy Half Space," *J. Res. NBS* 65D Radio Prop. No. 2, March-April, 1961.
[11] Maley and King, "Impedance of a Monopole Antenna With a Radial-Wire Ground System on an Imperfectly Conducting Half-Space, Part I," *J. Res. NBS* 66D Radio Prop. No. 2, March-April, 1962.
[12] Wait and Walters, "Influence of a Sector Ground Screen on the Field of a Vertical Antenna," NBS, Monograph 60, April 15, 1963.
[13] Maley and King, "Impedance of a Monopole Antenna With a Radial-Wire Ground System on an Imperfectly Conducting Half-Space, Part II," *J. Res. NBS* USNC-URSI, 68D, No. 2, February, 1964.
[14] Maley and King, "Impedance of a Monopole Antenna With a Radial-Wire Ground System on an Imperfectly Conducting Half-Space, Part III", *J. Res. NBS* USNC-URSI, 68D, No. 3, March, 1964.
[15] Gustafson, Chase and Balli, "Ground System Effect on High Frequency Antenna Propagation," *Research Report*, U.S. Navy Electronics Laboratory, January 4, 1966.
[16] Jager, "Effect of the Earth's Surface of Antenna Patterns in the Short Wave Range," *Int. Elek. Rundsch*, 24(4), pp. 101-104, (1970).
[17] Hill and Wait, "Calculated Pattern of a Vertical Antenna with a Finite Radial-Wire Ground System," *Radio Science*, Vol. 8, No. 1, pp. 81-86, January, 1973.
[18] Rafuse and Ruze, "Low Angle Radiation from Vertically Polarized Antenna Over Radially Heterogeneous Flat Ground," *Radio Science*, Vol. 10, No. 12, pp. 1011-1018, December, 1975.
[19] Stanley, "Optimum Ground Systems for Vertical Antennas," *QST*, December, 1976.
[20] Card, "Earth Resistivity and Geological Structure," *Electrical Engineering*, pp. 1153-1161, November, 1935.
[21] *Reference Data for Radio Engineers*, Fifth Edition, Howard W. Sams and Co., Inc., ITT, 26-3 to 26-5.
[22] Lagg, *Earth Resistances*, Pitman Publishing Corp. 1964, pp. 206-229.
[23] Private communication.
[24] See Ref. 1.

from August 1988 *QST*

Elevated Vertical Antenna Systems

Is your vertical-antenna system performance up to snuff? If not, maybe it needs a lift—in elevation above ground, that is!

By Al Christman, KB8I
Electrical Engineering Dept
Grove City College
Grove City, PA 16127

For many years, standard broadcast stations have used vertical monopoles (towers) as transmitting antennas. These monopoles are required by the FCC to have extensive ground systems, usually consisting of 120 or more buried radial wires that are used to simulate a perfectly conducting image plane beneath the monopole. The length of the radials is generally ¼ λ, although longer radials are often used. Electromagnetic energy leaving a radiator travels through space until reaching the earth's surface, where it flows through the soil to the radials, and then back to the antenna feed point.

Background

The FCC mandate requiring the use of many buried radials is apparently based upon the findings of three RCA engineers: Brown, Lewis and Epstein. These men carried out extensive tests on buried-wire radial ground systems in the mid-1930s and published their results in a now-classic paper in the *Proceedings of the Institute of Radio Engineers*.[1] In this 1937 paper, a single test was performed wherein the radials were laid upon the surface of the earth rather than buried in the soil. The conclusion was that "this ground system is about as good as an equal number of buried wires."[2] The experimenters' normal procedure was to bury the wires to a depth of about 6 inches.[3] Although this work was done at a frequency of 3 MHz, the results were quickly applied by AM broadcasters to their own part of the spectrum (540-1600 kHz), and buried radials have been used in AM-broadcast antenna systems ever since.

I recently studied elevated vertical antenna systems to determine how well they perform compared to conventional ground-mounted systems. My computer-modeling results indicate that an elevated vertical monopole antenna with four elevated horizontal radials provides more power gain at low elevation angles than does a conventional ground-mounted monopole with 120 buried radials.

The frequency of operation for my analyses was 3.8 MHz, and I used ground constants σ (conductivity) and ϵ_r (relative permittivity) that simulate average-soil electrical parameters. The computer program I used for this work was NEC-GSD, a *Method of Moments* code developed by engineers at the Lawrence Livermore National Laboratory.

In agreement with the findings of Arch Doty, K8CFU, I believe that the use of elevated, rather than buried, radials provides superior performance, because it allows the collection of electromagnetic energy in the form of *displacement currents*, rather than forcing *conduction currents* to flow through lossy earth.[4]

The Computer Analysis

The first step I took was to determine what effects, if any, would be caused by changing the depth at which the ground radials were buried. I used NEC-GSD to model a ¼-λ vertical monopole with 120 buried ¼-λ radials. The operating frequency was 3.8 MHz, and I modeled the system with average ground ($\sigma = 0.003$ S/m and $\epsilon_r = 13$).[5]

For the NEC model, the antenna was constructed of no. 12 wire (radius = 1 mm) and metal conductivities were adjusted to simulate an aluminum monopole mounted on a 2-foot steel ground stake with copper radials. As the burial depth of the radials was increased from 2 to 6 inches, the power gain of the antenna decreased only slightly (see Table 1), as did the ground-wave field strength. (Note that the reactive portion of the input impedance may be altered by adjusting the length of the monopole or by making it thicker in relation to the radials.) I used the vertical-monopole antenna system with 120 radials buried 2 inches deep as a reference standard for comparison with the other antenna systems discussed in this article.

I repeated the procedure described above using four buried radials, rather than 120. The results are given in Table 2. As before, slightly lower power gains and field strengths were calculated as radial depth increased. Compared to the 120-radial cases, monopoles with only four buried radials have much higher ground losses, as evidenced by

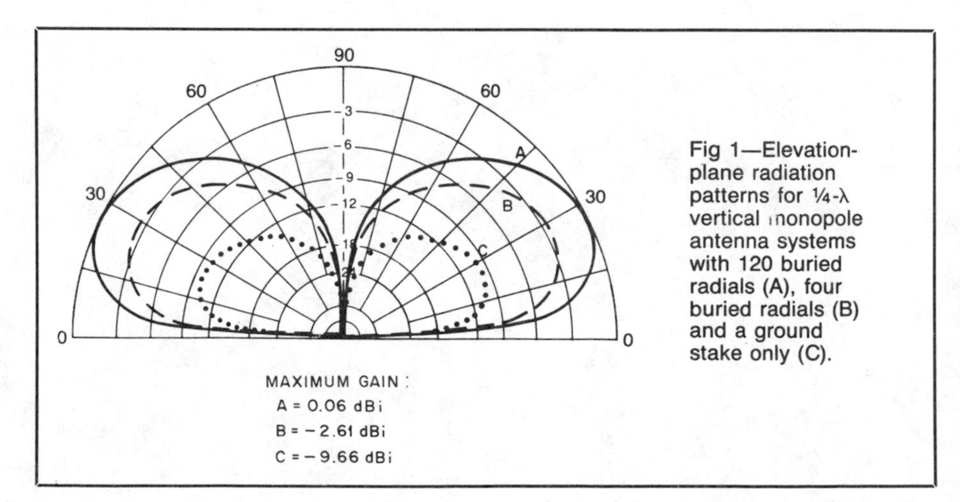

Fig 1—Elevation-plane radiation patterns for ¼-λ vertical monopole antenna systems with 120 buried radials (A), four buried radials (B) and a ground stake only (C).

Table 1

Power Gain and Electric Field Strength for Vertical Monopole Antennas with 120 Buried Radials

Elevation angle (degrees)	*Calculated power gain (dBi)* 120 radials buried 2 in.	120 radials buried 4 in.	120 radials buried 6 in.
0	−∞	−∞	−∞
5	−6.14	−6.15	−6.16
10	−2.40	−2.41	−2.42
15	−0.86	−0.87	−0.88
20	−0.17	−0.18	−0.19
25	+0.06	+0.04	+0.03
30	−0.02	−0.03	−0.04
40	−0.83	−0.84	−0.85
50	−2.37	−2.37	−2.37
60	−4.68	−4.69	−4.69
70	−8.13	−8.14	−8.14
80	−14.13	−14.14	−14.14
90	−158.38	−158.45	−158.51
Vertical electric field strength (mV/m)	33.16	33.10	33.06
Input impedance (ohms)	39.87 + *j*22.0	40.18 + *j*22.49	40.44 + *j*23.02

Table 2

Calculated Power Gain and Electric Field Strength for Vertical Monopole Antennas with 4 Buried Radials

Elevation angle (degrees)	*Calculated power gain (dBi)* 4 radials buried 2 in.	4 radials buried 4 in.	4 radials buried 6 in.
0	−∞	−∞	−∞
5	−8.82	−8.84	−8.85
10	−5.08	−5.10	−5.11
15	−3.54	−3.56	−3.58
20	−2.85	−2.87	−2.89
25	−2.62	−2.65	−2.66
30	−2.70	−2.72	−2.74
40	−3.52	−3.54	−3.55
50	−5.06	−5.08	−5.10
60	−7.37	−7.40	−7.42
70	−10.83	−10.86	−10.87
80	−16.84	−16.86	−16.88
90	−169.74	−169.99	−170.17
Vertical electric field strength (mV/m)	24.37	24.31	24.27
Input impedance (ohms)	74.48 + *j*33.69	74.73 + *j*34.04	74.93 + *j*34.39

Table 3

Calculated Power Gain and Electric Field Strength for Elevated Vertical Monopole Antenna Systems

Elevation angle (degrees)	*Calculated power gain (dBi)* 4 radials height = 5 ft	4 radials height = 10 ft	4 radials height = 15 ft	4 radials height = 20 ft	4 radials height = 25 ft	4 radials height = 30 ft
0	−∞	−∞	−∞	−∞	−∞	−∞
5	−6.40	−6.22	−6.09	−5.97	−5.82	−5.60
10	−2.69	−2.53	−2.43	−2.34	−2.23	−2.06
15	−1.19	−1.08	−1.03	−1.00	−0.96	−0.85
20	−0.56	−0.50	−0.53	−0.59	−0.64	−0.64
25	−0.41	−0.43	−0.54	−0.71	−0.89	−1.02
30	−0.57	−0.68	−0.91	−1.22	−1.56	−1.87
40	−1.59	−1.93	−2.45	−3.13	−3.92	−4.71
50	−3.38	−3.99	−4.88	−6.05	−7.46	−8.87
60	−5.94	−6.85	−8.15	−9.92	−12.06	−13.72
70	−9.61	−10.80	−12.50	−14.87	−17.55	−18.05
80	−15.77	−17.15	−19.16	−21.97	−24.67	−23.45
90	−157.37	−154.72	−152.32	−150.20	−148.37	−146.88
Vertical electric field strength (mV/m)	32.19	32.94	33.49	34.00	34.66	35.55
Input impedance (ohms)	38.64 + *j*8.60	36.06 + *j*3.37	3.77 + *j*0.59	31.35 − *j*1.17	28.82 − *j*2.05	26.51 − *j*2.08

reductions in gain and field strength, and by increases in input resistance. Much of the power radiated by the 120-radial antennas is now wasted heating the soil in the four-radial systems. Variations in azimuth-plane gain were negligible (0.01 dB or less).

The elevation-plane radiation patterns for vertical monopole antennas with 120 radials, four radials and no radials (ground stake only) are shown in Fig 1. The pattern shape remains essentially constant, but the pattern size (gain) depends upon the quality of the ground system. All of the vertical electric field strengths in the tables were normalized for an applied power of 1 kW at the feed points of the antennas, and were calculated at a distance of 1 mile and a height of 5 feet above ground. [This normalization procedure allows the pattern values for the different antenna configurations to be compared directly at any given elevation angle—Ed.] At this height the electric field is almost entirely surface wave (ground wave) rather than sky wave. I concluded from this comparison that if radials must be buried, more radials are better than a few radials in order to minimize losses. Also, it's helpful to keep buried radials as close as possible to the ground surface.

The next system configuration I modeled was a vertical monopole radiator with four horizontal radials—all elevated above the earth's surface. I found that low-angle power gain and field strength increase as the height of the antenna system is increased. Also, the gain at somewhat higher angles decreases as the antenna is raised (see Table 3). Note that the power gain at take-off angles below 15° increases continually as the antenna height is raised from 5 feet to 30 feet, but the gain at a take-off angle of 20° reaches a maximum value at a height of around 10 feet, and then decreases as the antenna is moved higher.

Compared with the reference 120-buried-radial system, the four-radial elevated antenna system reaches parity (equivalent performance to the reference) at low angles at a height of about 15 feet above the surface of the earth. If the antenna is raised further, a secondary high-angle lobe will develop as the antenna height approaches ¼ λ; gain at very low angles continues to increase. Variations in azimuth-plane gain are quite small, even with only four radials in the antenna system.

Antenna-System Geometry

The physical layout of the basic elevated-radial antenna system I modeled is shown in Fig 2. The monopole and the outer ends of the radials are supported by conductive masts. The height of the radial-support masts is the same as the elevation of the radials above ground, but these masts are separated laterally from the tips of the radials by six inches. The central mast supports the monopole, and it is also the same height as the radials. Each mast is attached to a two-foot-long ground stake that is driven fully into the earth.

The masts and ground stakes are made of steel, the radials are copper, and the monopole is constructed of aluminum, as before. All conductors are no. 12. The four radials are electrically connected directly to the top of the central mast, but are insulated from all other support structures. The outer conductor of the coaxial cable is also connected to the four radials at the top of the center mast.

This antenna arrangement allows significant current to flow on the central mast, so I also modeled the system with a slightly different feed-point design. In this second elevated-radial model, I isolated the central mast from the radials in order to reduce current flow on the mast.[6] As shown in Table 4, the isolated antenna system yields a small improvement in ground-wave field strength values, and more power gain at low elevation angles. There is only a slight change in feed-point impedance. Fig 3 shows the elevation-plane radiation pattern for an isolated four-radial

Table 4

Calculated Power Gain and Electric Field Strength for Isolated Elevated Vertical Monopole Antenna Systems

Elevation angle (degrees)	*Calculated power gain (dBi)* 4 radials height = 5 ft	4 radials height = 10 ft	4 radials height = 15 ft	4 radials height = 20 ft	4 radials height = 25 ft	4 radials height = 30 ft
0	−∞	−∞	−∞	−∞	−∞	−∞
5	−6.33	−6.01	−5.74	−5.50	−5.27	−5.06
10	−2.62	−2.32	−2.08	−1.87	−1.67	−1.50
15	−1.12	−0.87	−0.67	−0.51	−0.37	−0.26
20	−0.49	−0.30	−0.16	−0.07	−0.02	+0.01
25	−0.33	−0.21	−0.16	−0.17	−0.22	−0.32
30	−0.49	−0.46	−0.52	−0.64	−0.82	−1.08
40	−1.51	−1.70	−2.01	−2.44	−2.99	−3.66
50	−3.29	−3.74	−4.37	−5.20	−6.23	−7.48
60	−5.85	−6.58	−7.57	−8.85	−10.43	−12.21
70	−9.52	−10.50	−11.83	−13.55	−15.58	−17.30
80	−15.67	−16.84	−18.41	−20.44	−22.65	−23.65
90	−157.35	−154.72	−152.47	−150.61	−149.07	−147.80
Vertical electric field strength (mV/m)	32.47	33.71	34.82	35.86	36.86	37.82
Input impedance (ohms)	38.19 + *j*8.46	35.06 + *j*3.52	32.59 + *j*1.38	30.54 + *j*0.26	28.74 − *j*0.35	27.15 − *j*0.64

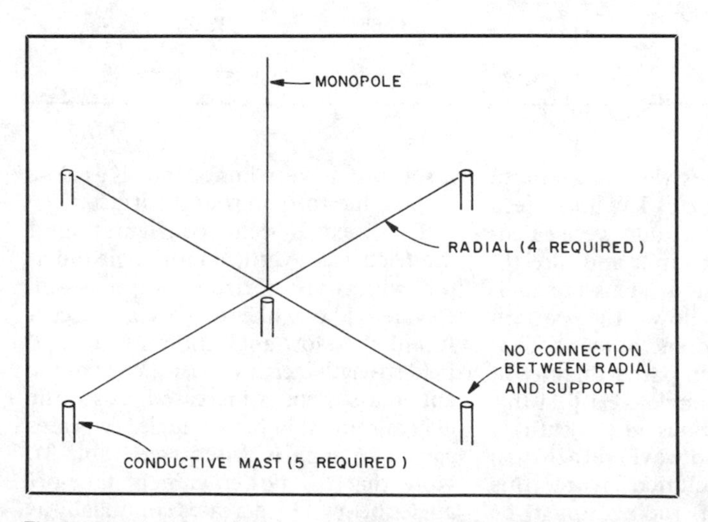

Fig 2—Physical layout of a four-radial, elevated vertical-monopole antenna.

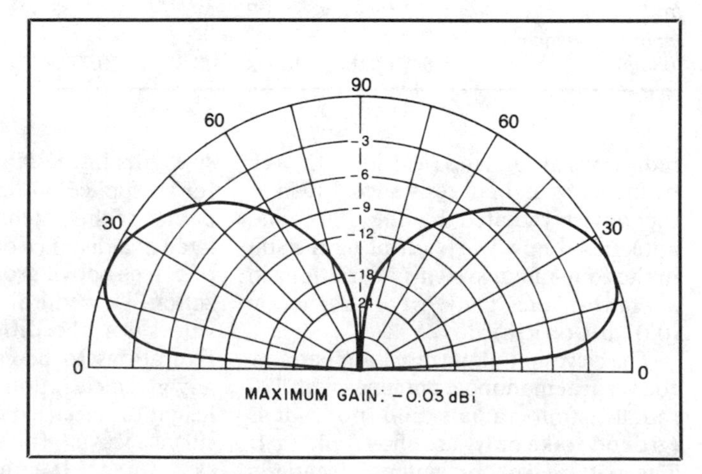

Fig 3—Elevation-plane radiation pattern for a four-radial, elevated vertical-monopole antenna isolated from the support mast (feed-point height = 15 feet).

Table 5

Calculated Power Gain and Electric Field Strength for Isolated Vertical Antenna Systems with 4 Shortened Radials and/or Shortened Monopoles

Elevation angle (degrees)	*Calculated power gain (dBi)* 1/8-λ monopole 1/8-λ radials	1/8-λ monopole 1/4-λ radials	1/4-λ monopole 1/8-λ radials
0	−∞	−∞	−∞
5	−6.88	−6.44	−6.20
10	−3.15	−2.74	−2.53
15	−1.62	−1.27	−1.09
20	−0.95	−0.67	−0.54
25	−0.74	−0.55	−0.49
30	−0.84	−0.77	−0.77
40	−1.71	−1.94	−2.09
50	−3.32	−3.93	−4.22
60	−5.72	−6.75	−7.14
70	−9.25	−10.66	−11.13
80	−15.31	−17.01	−17.51
90	−154.24	−146.52	−160.29
Vertical electric field strength (mV/m)	30.44	32.05	33.02
Input impedance (ohms)	8.25 −*j*653.45	7.0 −*j*541.76	36.32 −*j*136.61

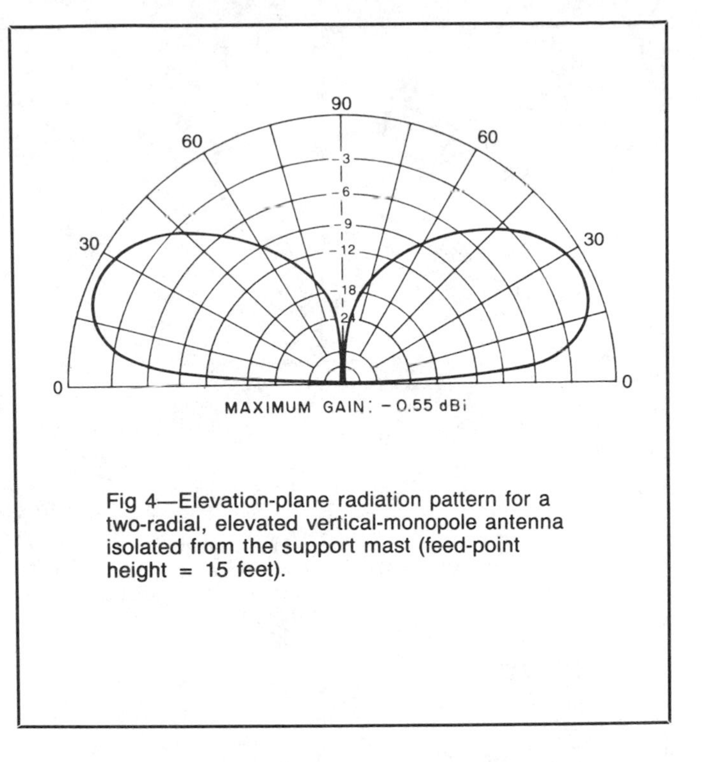

Fig 4—Elevation-plane radiation pattern for a two-radial, elevated vertical-monopole antenna isolated from the support mast (feed-point height = 15 feet).

Table 6

Calculated Power Gain and Electric Field Strength for Isolated Vertical Antenna Systems with 2 Radials

Elevation angle (degrees)	*Calculated power gain (dBi)* 1/8-λ monopole 1/8-λ radials, *Azimuth angle (degrees)* 0	45	90	1/8-λ monopole 1/4-λ radials, *Azimuth angle (degrees)* 0	45	90
0	−∞	−∞	−∞	−∞	−∞	−∞
5	−7.13	−7.05	−6.96	−7.75	−7.59	−7.32
10	−3.41	−3.31	−3.22	−4.09	−3.87	−3.58
15	−1.89	−1.78	−1.67	−2.68	−2.37	−2.04
20	−1.24	−1.11	−0.98	−2.17	−1.74	−1.35
25	−1.05	−0.90	−0.74	−2.18	−1.59	−1.11
30	−1.18	−0.99	−0.81	−2.54	−1.77	−1.18
40	−2.13	−1.86	−1.60	−4.09	−2.83	−1.98
50	−3.83	−3.46	−3.13	−6.52	−4.66	−3.51
60	−6.32	−5.85	−5.43	−9.75	−7.26	−5.83
70	−9.93	−9.37	−8.88	−13.96	−10.95	−9.28
80	−16.05	−15.42	−14.88	−20.41	−17.11	−15.29
90	−151.54	−151.54	−151.54	−144.30	−144.30	−144.30
Vertical electric field strength (mV/m)	29.61	29.87	30.16	27.59	27.96	28.88
Input impedance (ohms)	8.63 −*j*780.88			9.20 −*j*544.24		

antenna system at a height of 15 feet.

Modifications to the Basic Elevated-Radial System

As I've shown, the performance of the full-size, isolated, elevated antenna system consisting of a ¼-λ monopole and four ¼-λ radials at a height of 15 feet is competitive with a conventional 120-buried-radial, ground-mounted antenna. This section describes the results of modeling exercises I conducted with various combinations of shortened monopoles and/or radials. Table 5 shows what can be achieved with a mixture of 1/4- and 1/8-λ elements.

Perhaps surprisingly, the best performer of this group is a 1/4-λ monopole with four 1/8-λ radials. This configuration provides more signal strength than the other variations, and also has an input impedance closer to 50 Ω. [Note that this configuration provides the best performance of the *modified* configurations, but does not perform as well as a full-size elevated ¼-λ radiator with four ¼-λ elevated radials—Ed.]

To find out what impact on system effectiveness would be suffered by reducing the three-dimensional antenna system to two dimensions, I next modeled elevated-radial monopole antenna systems with only two radials. The results are shown in Tables 6 and 7. The ¼-λ monopole with two ¼-λ radials appears to be the best in this group, and is actually superior to the best of the four-radial "half-pints" previously described (in Table 5). The elevation-plane radiation pattern for this antenna configuration is shown in Fig 4.

Jack Belrose, VE2CV, suggested that I model some "Field-Day Special" antennas using an elevated monopole with just a single radial. The results are presented in Tables 8 and 9. These hybrids put out a mix of vertically and horizontally polarized radiation. They produce both low- and high-angle radiation, and exhibit front-to-back ratios as high as 12 to 15 dB at some takeoff angles. The full-size version (¼-λ elements) appears to work the best, and its feed-point impedance is much more favorable for 50-Ω feed lines than the rest of the

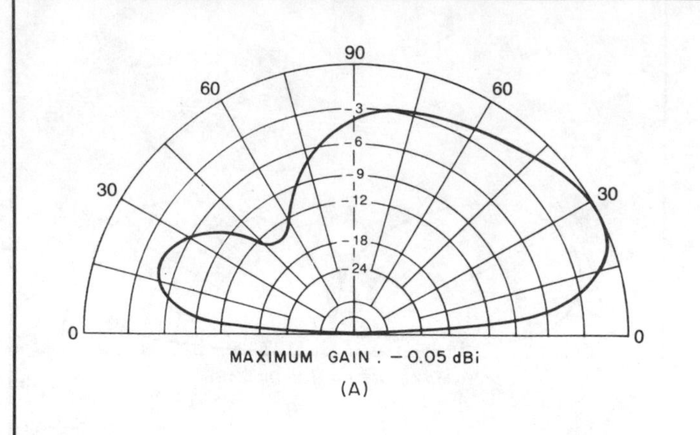

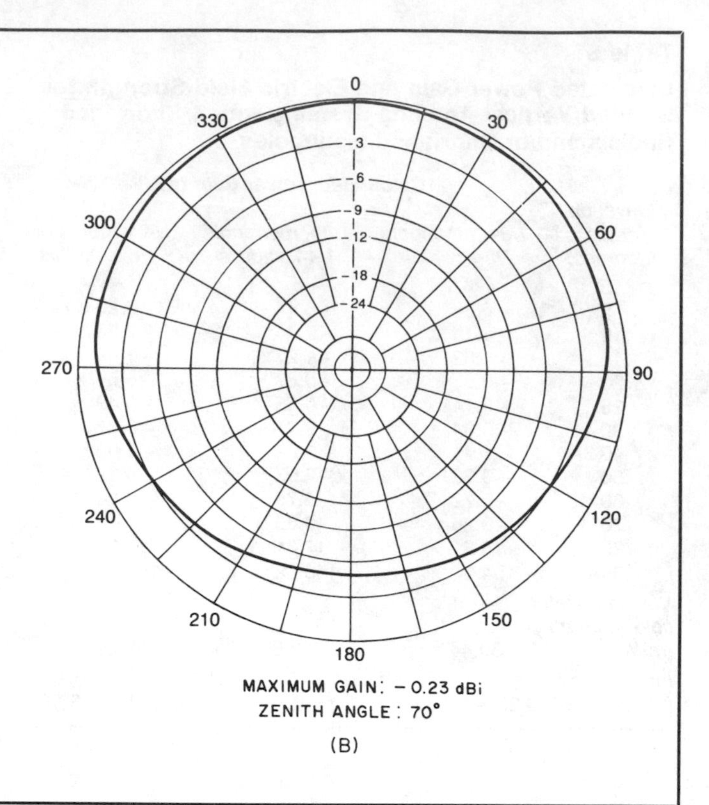

Fig 5—Elevation-plane (A) and azimuth-plane (B) radiation patterns for a one-radial, elevated vertical-monopole antenna isolated from the support mast (height = 15 feet).

bunch. Fig 5 shows the radiation patterns of this antenna.

Arrays

Many hams use phased-vertical arrays for 80-meter DXing, and elevated-radial antenna systems should lend themselves nicely to such applications. Table 10 lists power-gain and field-strength values for both the two-element end-fire (cardioid) array and the very popular four-square array, when constructed from individual four-elevated-radial building blocks. These antenna-system configurations are shown in Figs 6 and 7, and the patterns are shown in Figs 8 and 9.

Soil Types

As the electrical quality of the soil becomes worse, an elevated-radial antenna system must be raised progressively higher above the earth in order to reach performance on par with that of the reference 120-buried-radial vertical monopole. If the soil is very highly conductive, the reverse is true. At AM broadcast frequencies (1 MHz), my modeling studies have shown that adequate heights are 10 feet for very good soil, 16 feet for average soil and 23 feet for very poor soil.

This reveals another interesting aspect of using the elevated-radial technique: as the operating frequency decreases, the height at which system performance approaches that of the reference-standard system also increases for a given soil type. In other words, an elevated-radial 160-meter antenna would have to be higher above the earth than its 80-meter equivalent in order to obtain comparable performance over the same ground.

Table 7

Calculated Power Gain and Electric Field Strength for Isolated Vertical Antenna Systems with 2 Radials

Elevation angle (degrees)	*Calculated power gain (dBi)* 1/4-λ monopole, 1/8-λ radials — Azimuth angle (degrees) 0	45	90	1/4-λ monopole, 1/4-λ radials — Azimuth angle (degrees) 0	45	90
0	−∞	−∞	−∞	−∞	−∞	−∞
5	−6.29	−6.25	−6.21	−6.12	−6.03	−5.90
10	−2.61	−2.57	−2.53	−2.48	−2.36	−2.22
15	−1.18	−1.13	−1.09	−1.10	−0.95	−0.78
20	−0.64	−0.58	−0.53	−0.64	−0.43	−0.22
25	−0.60	−0.53	−0.46	−0.69	−0.42	−0.16
30	−0.90	−0.81	−0.73	−1.12	−0.77	−0.44
40	−2.26	−2.13	−2.00	−2.81	−2.22	−1.72
50	−4.43	−4.25	−4.07	−5.43	−4.53	−3.81
60	−7.41	−7.16	−6.93	−8.90	−7.63	−6.68
70	−11.45	−11.14	−10.86	−13.40	−11.78	−10.63
80	−17.87	−17.52	−17.19	−20.11	−18.26	−16.97
90	−157.65	−157.65	−157.65	−149.71	−149.71	−149.71
Vertical electric field strength (mV/m)	32.72	32.84	32.97	33.35	33.67	34.15
Input impedance (ohms)	36.81 −*j*263.26			34.92 +*j*0.83		

Summary

My studies on vertical monopole antennas using the NEC-GSD computer

Table 8

Calculated Power Gain and Electric Field Strength for Isolated Vertical Antenna Systems with 1 Radial

Elevation angle angle (degrees)	*Calculated power gain (dBi)*							
	1/4-λ monopole 1/4-λ radial				*1/4-λ monopole 1/8-λ radial*			
	Azimuth angle (degrees)				*Azimuth angle (degrees)*			
	0	*45*	*90*	*180*	*0*	*45*	*90*	*180*
0	−∞	−∞	−∞	−∞	−∞	−∞	−∞	−∞
5	−6.07	−6.25	−7.30	−9.52	−5.91	−6.08	−6.60	−7.53
10	−2.36	−2.52	−3.55	−5.96	−2.21	−2.38	−2.90	−3.88
15	−0.87	−1.00	−2.01	−4.75	−0.74	−0.91	−1.44	−2.49
20	−0.23	−0.33	−1.33	−4.54	−0.14	−0.30	−0.86	−2.02
25	−0.05	−0.11	−1.09	−4.97	−0.01	−0.17	−0.76	−2.08
30	−0.14	−0.16	−1.14	−5.92	−0.19	−0.36	−0.99	−2.51
40	−0.76	−0.71	−1.69	−9.36	−1.17	−1.36	−2.12	−4.24
50	−1.55	−1.49	−2.49	−13.82	−2.70	−2.93	−3.89	−7.02
60	−2.19	−2.20	−3.20	−11.80	−4.54	−4.85	−6.07	−10.72
70	−2.64	−2.77	−3.65	−7.82	−6.55	−6.94	−8.37	−14.10
80	−3.12	−3.28	−3.86	−5.35	−8.69	−9.08	−10.31	−13.70
90	−3.92	−3.92	−3.92	−3.92	−11.12	−11.12	−11.12	−11.12
Vertical electric field strength (mV/m)	33.21	32.43	28.69	22.90	33.98	33.35	31.49	28.52
Input impedance (ohms)	49.56 +*j*16.56				40.65 −*j*485.47			

code indicate that a radiator elevated 10 to 20 feet above ground and having only four elevated horizontal radials can outperform a ground-mounted monopole with 120 buried radials. At 3.8 MHz, an elevation height of about 15 feet is adequate for average soil, while a lower height is satisfactory for shorter wavelengths. Higher elevation above ground is necessary over soil with poorer electrical characteristics and at lower operating frequencies.

I will be doing field studies to verify the computer predictions (preliminary tests during Field Day showed very promising results). If the information gathered from NEC is correct, the construction cost and complexity of effective vertical-monopole antenna systems can be greatly reduced over that of comparable buried-radial systems now widely in use. At the same time, ease of installation and low-angle gain will be increased. The elevated-radial technique appears to be equally valid in the medium-frequency broadcast band and at the lower end of the HF range, so perhaps the ground-plane vertical is "the antenna for all bands"!

Table 9

Calculated Power Gain and Electric Field Strength for Isolated Vertical Antenna Systems with 1 Radial

Elevation angle (degrees)	*Calculated power gain (dBi)*							
	1/8-λ monopole 1/8-λ radial				*1/8-λ monopole 1/4-λ radial*			
	Azimuth angle (degrees)				*Azimuth angle (degrees)*			
	0	*45*	*90*	*180*	*0*	*45*	*90*	*180*
0	−∞	−∞	−∞	−∞	−∞	−∞	−∞	−∞
5	−8.54	−8.88	−10.84	−17.05	−6.96	−7.28	−8.33	−10.58
10	−4.76	−5.01	−6.79	−13.72	−3.19	−3.50	−4.54	−6.90
15	−3.13	−3.29	−4.88	−12.93	−1.61	−1.90	−2.92	−5.47
20	−2.31	−2.36	−3.75	−13.42	−0.85	−1.12	−2.14	−4.93
25	−1.86	−1.80	−3.00	−14.95	−0.52	−0.77	−1.79	−4.90
30	−1.63	−1.46	−2.48	−17.44	−0.46	−0.69	−1.71	−5.21
40	−1.41	−1.10	−1.82	−17.11	−0.83	−1.03	−2.08	−6.55
50	−1.19	−0.87	−1.42	−10.19	−1.58	−1.78	−2.88	−8.36
60	−0.88	−0.69	−1.16	−5.93	−2.52	−2.75	−3.89	−9.51
70	−0.59	−0.56	−0.99	−3.32	−3.56	−3.82	−4.91	−8.99
80	−0.52	−0.59	−0.89	−1.72	−4.68	−4.92	−5.68	−7.49
90	−0.85	−0.85	−0.85	−0.85	−5.97	−5.97	−5.97	−5.97
Vertical electric field strength (mV/m)	24.75	23.45	18.09	9.90	29.95	28.79	25.48	20.18
Input Impedance (ohms)	23.49 −*j*527.41				12.22 −*j*1004.27			

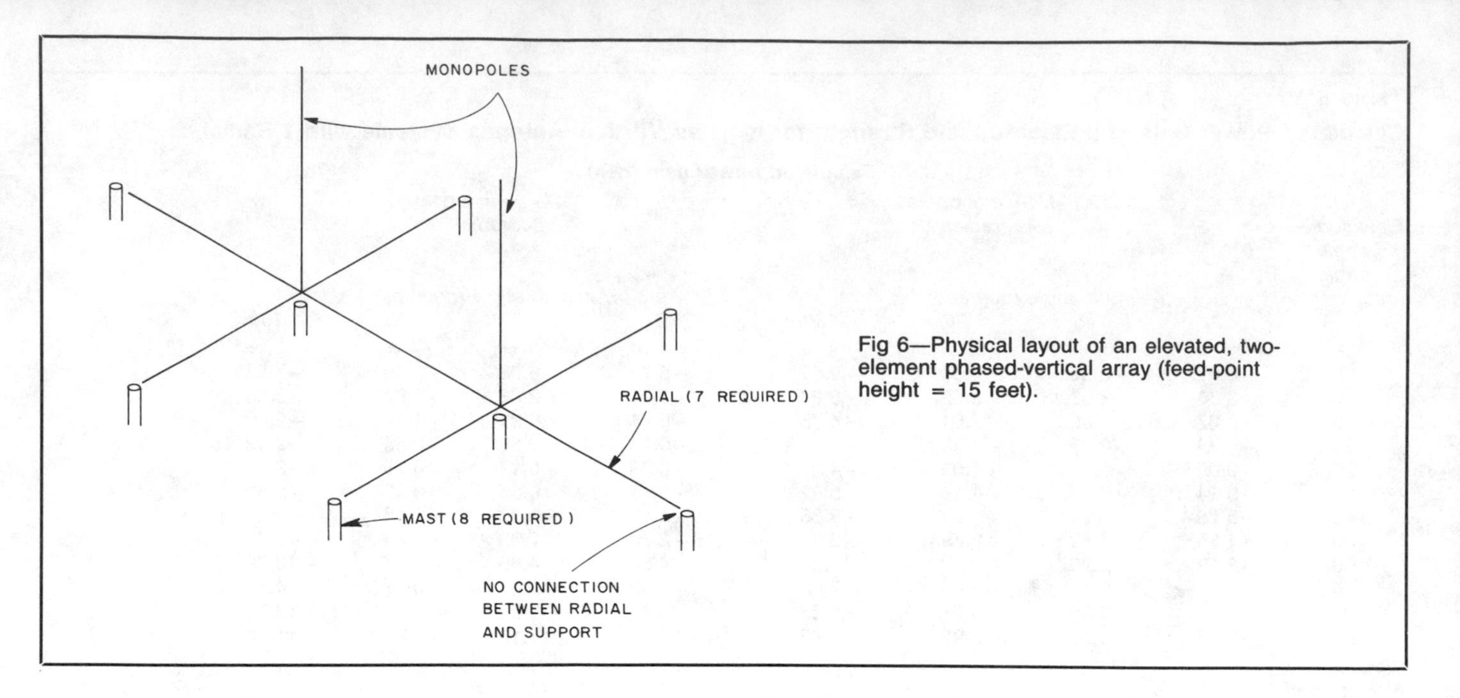

Fig 6—Physical layout of an elevated, two-element phased-vertical array (feed-point height = 15 feet).

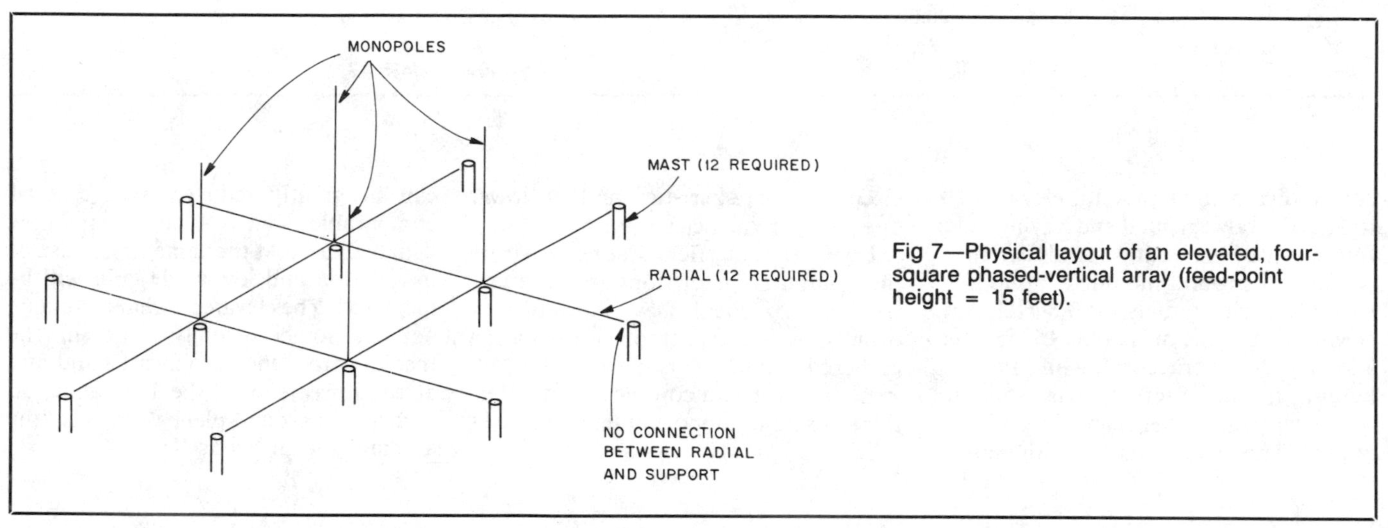

Fig 7—Physical layout of an elevated, four-square phased-vertical array (feed-point height = 15 feet).

Table 10

Calculated Power Gain and Electric Field Strength for Isolated Phased-Vertical Antenna Systems

	Calculated power gain (dBi)							
	2-element cardioid array				*4-element 4-square array*			
Elevation angle (degrees)	*Azimuth angle (degrees)*				*Azimuth angle (degrees)*			
	0	*45*	*90*	*180*	*45*	*90*	*135*	*225*
0	−∞	−∞	−∞	−∞	−∞	−∞	−∞	−∞
5	−2.45	−2.96	−6.18	−32.85	−0.50	−3.29	−13.87	−37.32
10	+1.23	+0.69	−2.53	−27.61	+3.16	+0.39	−9.94	−33.19
15	2.66	2.07	−1.16	−23.74	4.54	1.82	−8.15	−31.06
20	3.20	2.55	−0.70	−20.42	5.03	2.36	−7.17	−29.65
25	3.23	2.51	−0.76	−17.68	5.00	2.40	−6.61	−28.67
30	2.92	2.10	−1.20	−15.51	4.63	2.11	−6.33	−28.00
40	+1.51	+0.47	−2.91	−12.70	3.17	+0.85	−6.33	−26.44
50	−0.76	−2.06	−5.57	−11.57	+1.08	−0.94	−6.71	−22.02
60	−3.87	−5.40	−9.11	−11.93	−1.29	−2.89	−7.10	−16.70
70	−8.01	−9.69	−13.60	−13.99	−3.61	−4.66	−7.35	−12.58
80	−14.06	−15.65	−19.44	−19.06	−5.65	−6.17	−7.48	−9.66
90	−25.00	−25.00	−25.00	−25.00	−7.53	−7.53	−7.53	−7.53
Vertical electric field strength (mV/m)	51.39	48.35	32.99	1.89	64.86	46.34	12.90	0.81
Input impedance (ohms)	19.61	+*j*7.41	51.07	+*j*33.71	9.68 +*j*2.78	36.8 −*j*4.47	36.8 −*j*4.47	66.62 +*j*47.2

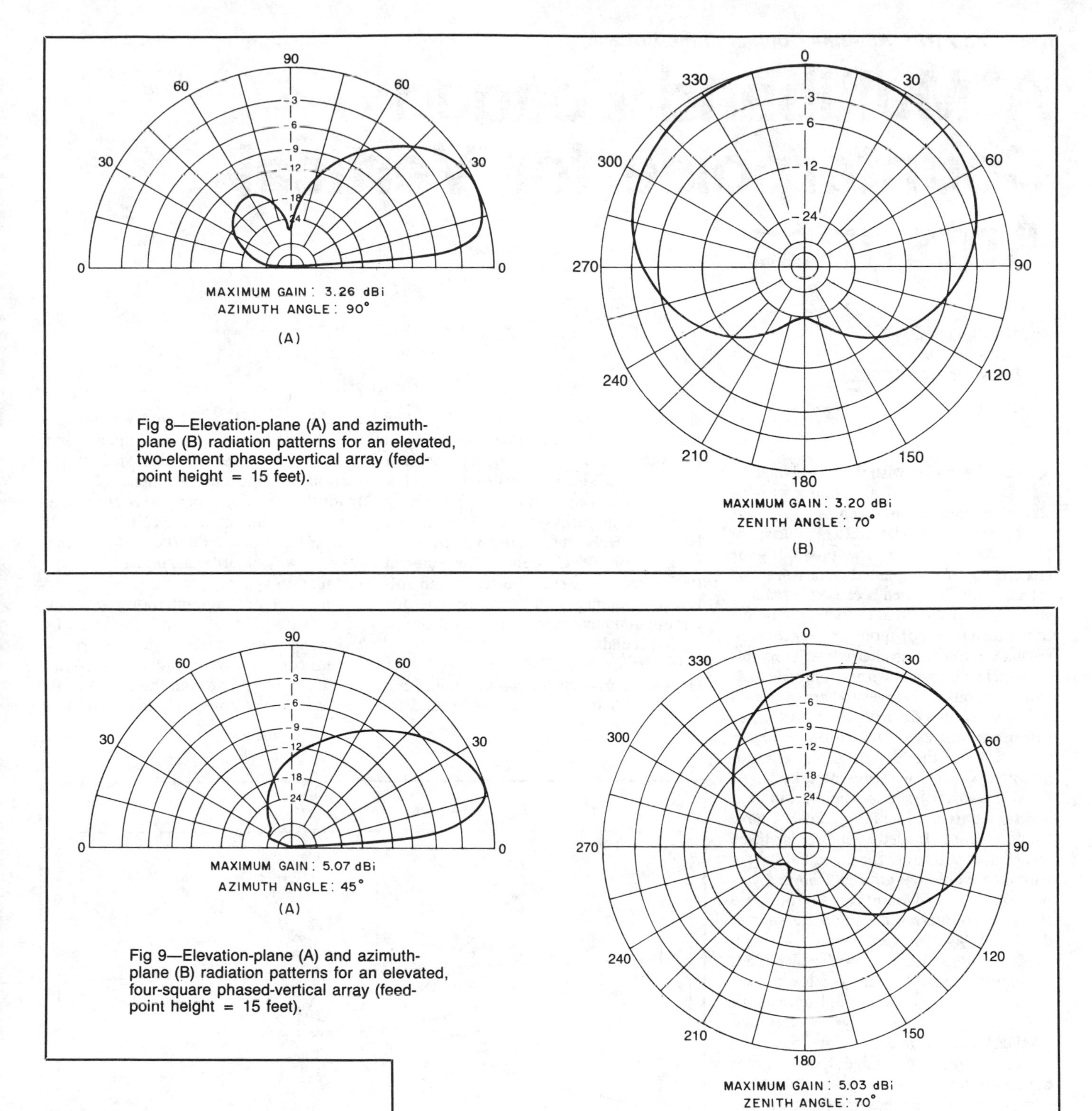

Fig 8—Elevation-plane (A) and azimuth-plane (B) radiation patterns for an elevated, two-element phased-vertical array (feed-point height = 15 feet).

Fig 9—Elevation-plane (A) and azimuth-plane (B) radiation patterns for an elevated, four-square phased-vertical array (feed-point height = 15 feet).

Notes

[1]G. H. Brown, R. F. Lewis and J. Epstein, "Ground Systems as a Factor in Antenna Efficiency," *Proceedings of the Institute of Radio Engineers*, Vol 25, No. 6, Jun 1937, pp 753-787.

[2]"Ground Systems...," p 784.

[3]"Ground Systems...," p 769.

[4]A. Doty, "Improving Vertical Antenna Efficiency," *CQ*, Apr 1984, pp 24-31.

[5]G. Hagn, SRI International, Arlington, VA, private communication.

[6]Because the coaxial transmission line feeding power to the antenna must extend vertically along the center mast, some means of preventing the flow of antenna current on the outer surface of the cable shield must be used to isolate the feed line from the vertical support structure. This can be done by using a transformer at the feed point, or by placing suitable ferrite material around the outside of the transmission line (a choke balun).

Reference

A. M. Christman, "Vertical Monopoles with Elevated Ground Systems," *Proceedings of the Third Annual Review of Progress in Applied Computational Electromagnetics*, Naval Postgraduate School, Monterey, CA, March 1987.

from *The ARRL Antenna Compendium, Vol 2*

A Multiband Loaded Counterpoise for Vertical Antennas

By H. L. Ley, Jr, N3CDR
PO Box 2047
Rockville, MD 20852

Did you ever wish you could have an efficient vertical antenna without installing "umpteen" radials? For most amateurs with limited available space for a radial system, the possibility of erecting a ground-mounted vertical antenna is frequently dismissed because ¼-λ radials will not fit in the usual residential city lot. Many turn to a roof-mounted vertical with resonant radials as an alternative. True, this is an efficient system with a number of advantages, but it removes the antenna to a point where it is difficult to reach for adjustments and experimentation. Another approach to the design of a ground-mounted vertical was presented by L. A. Moxon, G6XN.[1] Moxon points out that several amateurs, including himself, have found a tuned, loaded counterpoise to be an efficient substitute for the radial system with a conventional ground-mounted vertical radiator. Moxon makes a strong case that the counterpoise should be shorter than ¼ λ, as it normally would be for a roof-mounted vertical with insulated, resonant radials.[2] He sets as a lower limit of total physical counterpoise length about 1/8 λ, or approximately 18 feet for 7 MHz.[3] These concepts led me to explore the development of a tuned, loaded multiband counterpoise for a ground-mounted vertical antenna that has demonstrated on-the-air performance equivalent to a radial system of ten radials up to 37 feet long.

The counterpoise system is made from two 8-foot lengths of ¾-inch aluminum tubing and the associated loading inductance. These extend 8½ feet horizontally at either side of the base of the vertical antenna. The visual impact is acceptable, and the space requirements are so modest that the system will fit into most city residential lots.

Mechanical Details

I chose a Hustler 4-BTV antenna for the installation because it was available. Other multiband or single-band vertical antennas should give comparable results. The antenna was ground-mounted according to the manufacturer's instructions with a 4-foot long, 2-inch diam steel pipe, sunk 2½ feet into the ground in the center of a 48 × 76 foot backyard area. The antenna base was insulated from the pipe by a section of split PVC tubing. The antenna tubing sections were adjusted, again following the manufacturer's directions, for a ground-mounted antenna with surface or buried radials.

In the counterpoise, the two 8-foot lengths of ¾-inch aluminum tubing are supported 8 inches above the ground by ceramic standoff insulators on wooden stakes. The tubing sections are placed in a straight line, one on either side of the vertical, with the inner ends spaced 6 inches from the antenna. They are connected at their inner ends by a no. 12 wire jumper attached at the center to the loading inductance. Every effort was made to keep the installation symmetrical.

Operation of the counterpoise was compared with that of a radial system by switching between the two. The radial system consisted of 10 radial wires, symmetrically placed around the antenna. At the base of the antenna, each radial was

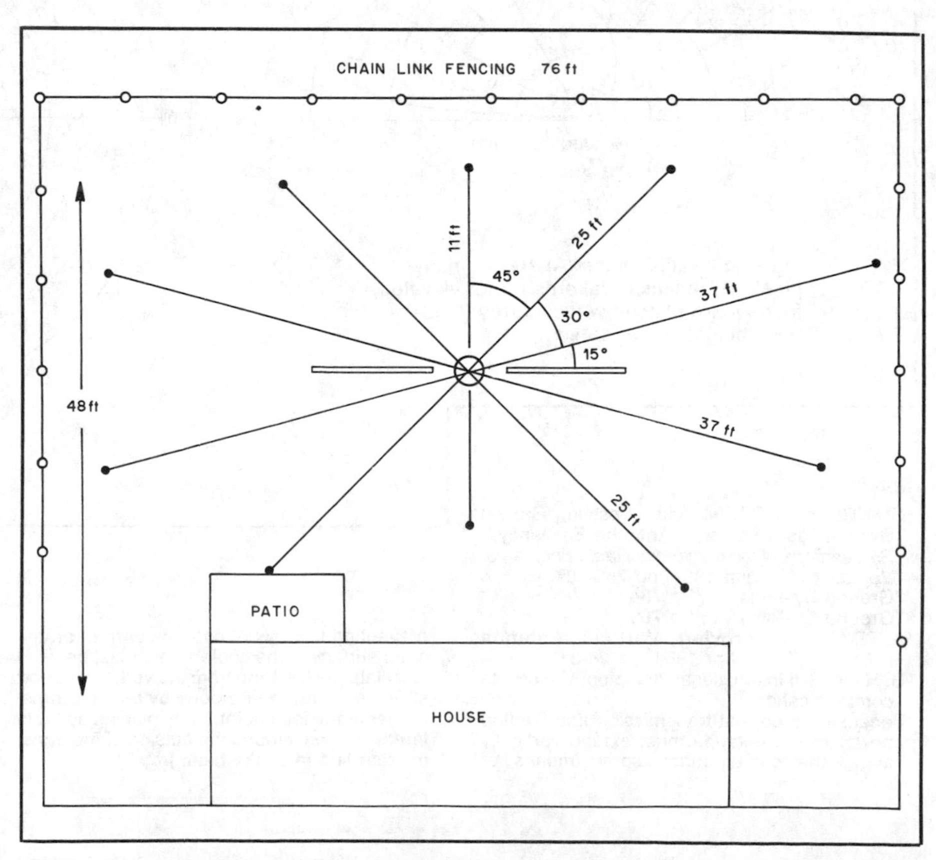

Fig 1—An overhead view of the antenna, radials, counterpoise, and surrounding area. The counterpoise is made of aluminum tubing and is perpendicular to the 11-foot-long radials.

connected with a solder lug to a stainless-steel disk, drilled and tapped for no. 8-32 machine screws. The radials were cut from no. 14 copper wire for the maximum length permitted by the size of the lot. Four of the radials were 37 feet, four were 25 feet, and two were 11 feet long. Each radial was pinned to the ground with 5 inch U-shaped "staples" cut and formed from coat hangers. The installation is outlined in Fig 1. In order to achieve symmetry in the radial installation, opposing radials were laid in a straight line, passing through the axis of the antenna. Originally, I gave thought to attaching the radials to the chain link fence surrounding the yard on three sides, but this was dismissed for reasons of loss of symmetry and of possible harmonic interference generated at rusty fence joints.

Electrical Details

For initial tests and evaluation of the vertical and radial system, an electrical ½ λ of 50-Ω RG-58 cable at 7.125 MHz was used with a Palomar noise bridge and appropriate low-reactance resistors installed in PL-259 plugs.[4] Initially, the bridge was set to zero on the reactance scale and balanced by adjustment of the station receiver frequency and the resistance potentiometer in the bridge. Each reading was repeated three times to permit averaging the result. The results are presented in Table 1. The measured antenna resistance varied between 45 and 70 ohms at average resonant frequencies of 7.10, 14.08, 21.40 and 28.27 MHz. These results were "distorted" by the fact that the feed line was not an exact multiple of ½-λ on all frequencies, although it was close in the 40- and 15-meter bands. Nevertheless, the results indicated that the antenna and radial system would give a low SWR with 50-Ω cable on all bands, a fact that was confirmed in subsequent on-the-air testing.

The antenna tubing section lengths could have been adjusted on 14, 21 and 28 MHz to bring the resonant frequency closer to the center of the bands, but I decided at the beginning of this evaluation to use the lengths recommended by the manufacturer in all subsequent measurements and testing.

Next, a 2-inch diameter, 2½-inch long loading coil of 8 turns per inch was selected from the junk box (such as B&W no. 3900 or AirDux no. 1608T). Alternate turns of the coil were bent in toward the center on one side of the coil to permit tapping it at 1.5, 3.5, 5.5 turns and so on. The starting turn could be tapped easily from the end of the coil. This inductance was connected from the coax braid connection on the insulated antenna mount through a relay to the counterpoise with small alligator clips. A 12-V 4-pole double-throw relay (Radio Shack No. 275-214) with all four similar contacts connected in parallel was wired to permit remote switching between the radial system and loaded counterpoise. A schematic of the installation is shown in Fig 2.

The station receiver, noise bridge and resistance standards were again put to work to determine the appropriate taps on the loading inductance for 40, 20, 15 and 10-meter operation with the counterpoise, using two separate sections of 50-Ω coax cut for ½ λ at 7.05 and 7.125 MHz. Results of these measurements are shown in Table 2. It is apparent that the loading of the

Table 1
Noise Bridge Data on Antenna and Radial System
Zero reactance was measured at all frequencies in this table.

Band	*Resistance, Ohms*	*Freq, MHz*
40 m	45	7.109
	50	7.110
	45	7.081
	Average 47	7.10
20 m	60	14.100
	50	14.019
	60	14.123
	Average 57	14.08
15 m	65	21.430
	65	21.378
	70	21.380
	Average 67	21.40
10 m	60	28.250
	65	28.276
	60	28.284
	Average 62	28.27

Measurements were made through a 7.125-MHz ½-λ section of RG-58 cable with Palomar noise bridge.

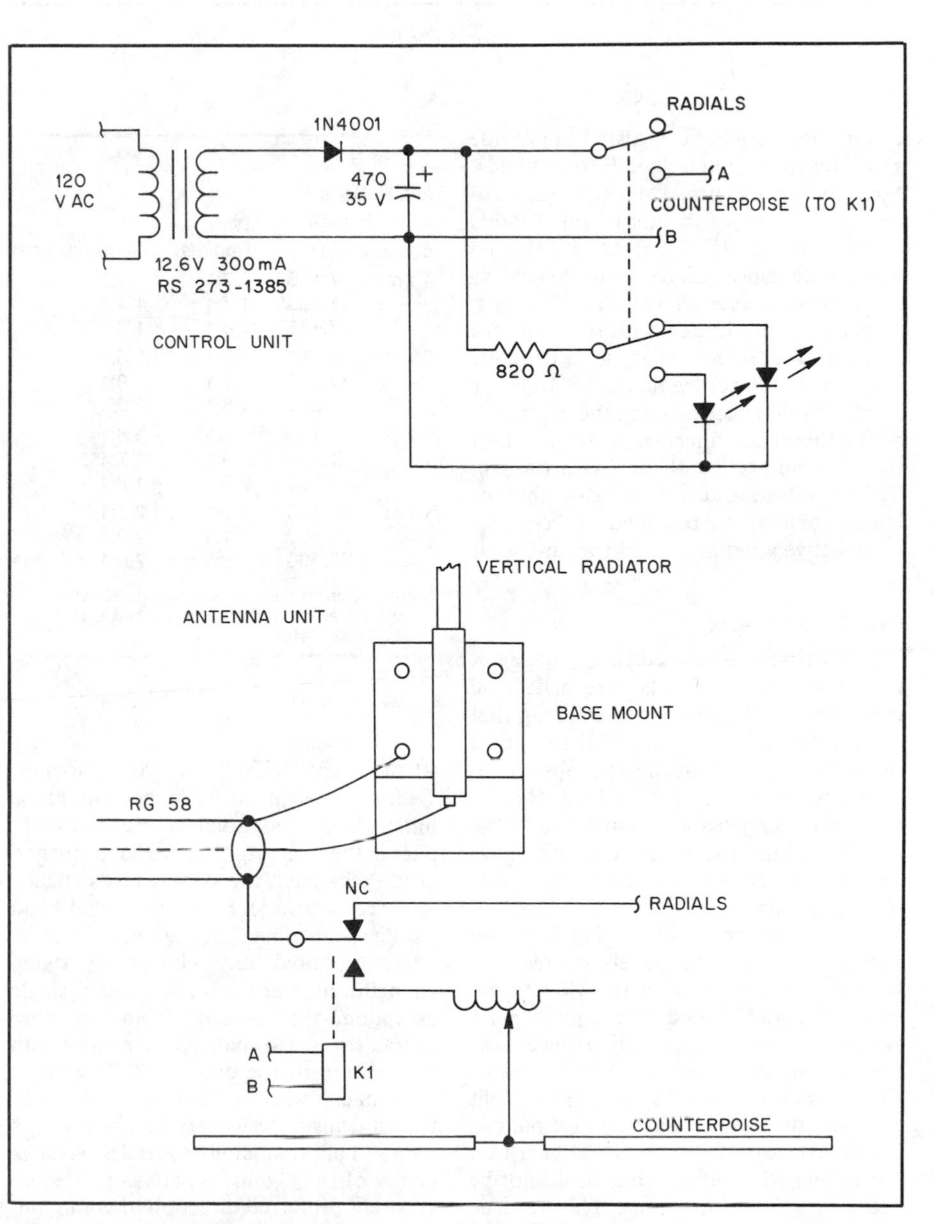

Fig 2—The remote switching arrangement used to switch between the counterpoise and ground radial system from the operating position.
K1—Radio Shack® four-pole double-throw relay, part no. RS275-214.

Table 2
Noise Bridge Data on Antenna and Counterpoise System

RG-58 cables were cut to ½ λ at 7.050 and 7.125 MHz.

Band	½λ Cable MHz	Measurement Freq, MHz	Loading Coil Tap, Turns	Resistance, Ohms	Reactance Dial
40 m	7.125	7.125	11.5	55	L 10
			9.5	40	C 2
			7.5	60	C 10
40 m	7.050	7.050	11.5	55	L 10
			9.5	45	0
			7.5	50	C 10
20 m	7.050	14.100	5.5	125	L 10
			3.5	50	0
			1.5	55	C 10
15 m	7.050	21.150	3.5	100	L 20
			1.5	75	0
			0.5	70	C 10
			1.5	70	0
10 m	7.050	28.000	None	60	C 10
		28.390	None	60	0

counterpoise could be adjusted to provide a nonreactive match in all four bands. Measured resistance values at a zero-reactance setting of the noise bridge varied between 45 and 75 ohms. At 28 MHz no added inductance was necessary to achieve system resonance at 28.390 MHz. This suggests that it may be desirable to extend the counterpoise, if the system is to be made resonant at a lower frequency on 10 m, by increasing the space between the inner ends of the tubing and increasing the length of the wire connecting them. With the collection of these results, it was clear that the loaded counterpoise provided an acceptable nonreactive antenna, ready for on-the-air testing.

Table 3
SWR Data

Band	Freq, MHz	With Radials	With Counterpoise
40 m	7.105	1.7:1	1.1:1
	7.125	1.7:1	1.1:1
	7.145	1.8:1	1.2:1
20 m	14.050	1.3:1	1.5:1
	14.150	1.7:1	1.8:1
	14.200	1.8:1	2.0:1
15 m	21.120	1.8:1	1.8:1
	21.150	2.0:1	1.7:1
	21.180	2.0:1	1.8:1
10 m	28.100	2.0:1	2.1:1
	28.300	2.1:1	2.0:1
	28.400	2.2:1	2.0:1

Measurements were made with a 25-watt-output transmitter and Heath HM-9 HF SWR/wattmeter.

On-the-Air Testing

Testing was conducted in three stages. First, SWR measurements were made in all four bands with the vertical working first against the radial system, and then against the counterpoise with the appropriate inductance connected in the circuit. Results of these tests are presented in Table 3. The radials and the loaded counterpoise gave comparable results in terms of SWR measurements.

The second stage of testing included checking the S-meter levels of received signals as the antenna was switched between the radial and the counterpoise system. No consistent difference was observed at levels above S-6. At levels of S-3 or lower, the radial system had a slight edge over the counterpoise, up to about one S unit. At no time was a signal heard on the system with radials that could not be heard with the counterpoise. This was true on all four bands.

The third stage of testing was use of the antenna in actual contacts, mainly on 40 m in the 7.100-7.150 MHz Novice/Technician segment. Contact would be initiated with the radials in use, and then, later in the contact, after RST reports had been exchanged, the system was switched to the counterpoise. I announced that I had "switched antennas" and asked if the other operator noted any change in signal strength. In every case, without a single exception, the response from the other operator was "no change," sometimes with a repetition of the original RST report.

Contacts with a 25-watt crystal-controlled transmitter covered a circle with a radius of approximately 400 miles. A quick review of the log and an atlas revealed no apparent preferred direction of communication of the counterpoise system. Further testing may reveal some favored direction for the counterpoise, but it is not apparent at this point in evaluation of the system.

Discussion and Conclusions

I approached this project with considerable skepticism, despite the convincing arguments of Moxon that a vertical monopole with loaded counterpoise was electrically the near equivalent of a vertical dipole. The test results demonstrate that Moxon was correct, and, further, that the loaded counterpoise is equivalent to a modest system of ten radials in on-the-air testing.

Obviously, the Hustler 4-BTV antenna requires some "fine tuning" to adjust its element lengths for resonance in the desired portions of the bands covered. Do not be discouraged if the loaded counterpoise reduces the claimed bandwidth of your vertical antenna. I remind you that the loaded counterpoise will reduce the bandwidth of this or any other vertical antenna because of the inclusion of another tuned element in the system.

I would like to have tested the system in actual contacts on bands other than 40 m before preparing this paper, but because of time, license and propagation limitations was unable to do so. Performance on 40 m was considered the most critical test of the concept because the counterpoise length was approximately 1/8 λ at 7 MHz. Others can fill in the information for 20, 15 and 10 m, where the counterpoise is a greater fraction of a wavelength.

Some readers may question the use of only 10 radials for comparison with the counterpoise, pointing out that the efficiency of a vertical radiator is reduced if the ground resistance of the radial system is too high.[5] The radial system included four elements greater than 1/4 λ at 7 MHz, and another four at approximately 1/5 λ. Although this does not meet the commercial broadcast standard of 120 radials, it is probably a better radial system than used by most amateurs, and is certainly as good or as better than the usual rooftop insulated, resonant radial system. It is possible, as noted in *The ARRL Antenna Book*, that a system of twenty 1/8-λ radials might have been preferable to the system used here. But that would have doubled the number of wires that could become entangled in the lawn mower, an environmental hazard of some concern! Should you explore the use of a larger number of radials, I would be interested to learn the results.

There remains one unanswered question in my mind at this point in the project. To what degree are the favorable results observed with the counterpoise system attributable to the presence of an isolated radial ground screen below it? I have no answer to this question except to believe that the influence of the radials on the counterpoise is minimal because of another example given by Moxon in his book.[6] In his Fig 11.16, Moxon sketches a monopole with a short inductively loaded counterpoise for 20 m. In this instance the counterpoise loading is not lumped but rather distributed, with the 21-foot counterpoise bent roughly into the shape of a trombone at the base of the vertical ¼-λ monopole element. He points out that this antenna may be mounted at ground level, but that improvement in performance "of at least half an S-unit can be expected if the base height is raised to about 0.2 λ, or 14 feet at 14 MHz." Moxon's antenna had no radial system below it, and, except for the type of loading used, does not differ greatly from the design presented here.

Elevation of the base of the antenna is attractive for other reasons, namely reduced earth losses and environmental safety. On the other hand, it just is not always feasible to elevate the base of the vertical element by 0.2 λ, or 28 feet, at 40 m. In my installation, there are no young children in the household or immediate neighborhood, so the chance of accidental contact of persons with the counterpoise element is minimal in a fenced backyard. A potential hazard of RF burns exists with any ground-mounted vertical element, even when operated against grounded radials, so the counterpoise is not considered to pose a significantly greater hazard than a ground-mounted vertical. The advantages of having the antenna feed point easily accessible for adjustment of the counterpoise loading inductances is obvious, and even critical, if the taps are selected by an electrically or manually operated switch. If I operated high power, I would give thought to extending and guying the mounting pipe so the antenna and counterpoise might be mounted about 6 feet above the ground. This would significantly improve environmental safety, yet still permit access by ladder for adjustment or repair.

As yet, I have not found a mathematical method to calculate the inductance required to resonate the 17-foot center-fed counterpoise, as I did with a loaded short horizontal dipole.[7]

The data given here will provide a starting point if you are interested in installing a similar system. The values of inductance used for this ground-mounted vertical with loaded counterpoise are 4.3, 0.92 and 0.52 μH for 40, 20 and 15 m respectively. I will be pleased to hear from others installing a similar system, and will reply to comments or questions if a stamped return envelope is enclosed with the letter. So, good luck, down with radials, and up with the counterpoise!

Notes

[1]L. A. Moxon, *HF Antennas for All Locations* (RSGB: Potters Bar, Herts, 1982), pp 4, 154-157, 164-165.

[2]See Moxon (note 1), pp 144-145.

[3]See Moxon (note 1), p 155.

[4]B. S. Hale, Ed., *The ARRL Handbook for the Radio Amateur*, 66th ed. (Newington: ARRL, 1989), p 25-23.

[5]G. L. Hall, Ed., *The ARRL Antenna Book*, 15th ed. (Newington: ARRL, 1988), pp 2-33, 2-34.

[6]See Moxon (note 1), pp 164-165.

[7]H. L. Ley, Jr, "Short Loaded Half-Wave Dipole Design—The Easy Way," *The ARRL Antenna Compendium, Volume 1* (Newington: ARRL, 1985), pp 116-122.

Antenna Products Suppliers

ARRL maintains an online resource to help locate suppliers for all your Amateur Radio needs. Point your Web browser to **http://www.arrl.org/tis/tisfind.html**. This page will allow you to search the ARRL TIS (Technical Information Service) database of more than 2000 suppliers who can provide goods and services of interest to radio amateurs. These include manufacturers, dealers, publications, clubs, museums, and so on. Or you can browse a list of categories of products and services. You may also download **tisfinst.exe** to use this database off-line.

You may search for company name, product, in fact any word contained in the record. Most common searches are for addresses, telephone numbers and e-mail addresses. You can enter a company name (full or partial), or to search under product/services you can enter a KEY WORD to receive a list of all records with that KEY WORD.

This list of categories (with the number of entries as this is written in November 1999) is for suppliers under the general category of ANTENNA:

- Aluminum Tubing (9)
- Antenna Dealer (67)
- Antenna Manufacturer (99)
- Antenna Part (55)
- Antenna Relay (13)
- Antenna Switch (33)
- Antenna Tuner (38)
- Balun (53)
- Beam Heading Chart (3)
- Coax (69)
- Coax Sealant (1)
- Diplexer (13)
- Discone Antenna (2)
- Dish Antenna (9)
- Dummy Load (33)
- Duplexer (32)
- Fiberglass Material (13)
- Gin Pole (7)
- Ground Plane Antenna (2)
- Guy Wire (25)
- J Pole Antenna (15)
- Ladder line (23)
- Log Periodic Antenna (1)
- Loop Antenna (24)
- Mast (29)
- Microwave Antenna (13)
- Mobile Antenna (62)
- Quad Antenna (18)
- Rotator (41)
- Rotator Control Cable (3)
- Satellite Antenna (19)
- Stacking Hardware (15)
- Tower (37)
- Tower Hardware (12)
- Trap (13)
- Tuner (1)
- Twin Lead (13)
- Twinlead (8)
- Waveguide (4)
- Wire (2)
- Yagi Antenna (37)

The advantage of using this online resource is that it is kept up to date by ARRL Headquarters staff. Be sure to check out the other services available on *ARRLWeb*. These inlcude ARRL advertiser links at **http://www2.arrl.org/ads/adlinks/html**. There is also a members-only section that includes a *QST* Product Review archive and a search function for *QST* and *QEX*. If you have not visited the members-only section, you will be pleased with the exclusive information and services available there to ARRL members.

Notes

Notes

Notes

About the ARRL

The seed for Amateur Radio was planted in the 1890s, when Guglielmo Marconi began his experiments in wireless telegraphy. Soon he was joined by dozens, then hundreds, of others who were enthusiastic about sending and receiving messages through the air—some with a commercial interest, but others solely out of a love for this new communications medium. The United States government began licensing Amateur Radio operators in 1912.

By 1914, there were thousands of Amateur Radio operators—hams—in the United States. Hiram Percy Maxim, a leading Hartford, Connecticut, inventor and industrialist saw the need for an organization to band together this fledgling group of radio experimenters. In May 1914 he founded the American Radio Relay League (ARRL) to meet that need.

Today ARRL, with approximately 170,000 members, is the largest organization of radio amateurs in the United States. The ARRL is a not-for-profit organization that:

- promotes interest in Amateur Radio communications and experimentation
- represents US radio amateurs in legislative matters, and
- maintains fraternalism and a high standard of conduct among Amateur Radio operators.

At ARRL headquarters in the Hartford suburb of Newington, the staff helps serve the needs of members. ARRL is also International Secretariat for the International Amateur Radio Union, which is made up of similar societies in 150 countries around the world.

ARRL publishes the monthly journal *QST*, as well as newsletters and many publications covering all aspects of Amateur Radio. Its headquarters station, W1AW, transmits bulletins of interest to radio amateurs and Morse code practice sessions. The ARRL also coordinates an extensive field organization, which includes volunteers who provide technical information and other support for radio amateurs as well as communications for public-service activities. ARRL also represents US amateurs with the Federal Communications Commission and other government agencies in the US and abroad.

Membership in ARRL means much more than receiving *QST* each month. In addition to the services already described, ARRL offers membership services on a personal level, such as the ARRL Volunteer Examiner Coordinator Program and a QSL bureau.

Full ARRL membership (available only to licensed radio amateurs) gives you a voice in how the affairs of the organization are governed. ARRL policy is set by a Board of Directors (one from each of 15 Divisions). Each year, one-third of the ARRL Board of Directors stands for election by the full members they represent. The day-to-day operation of ARRL HQ is managed by an Executive Vice President and a Chief Financial Officer.

No matter what aspect of Amateur Radio attracts you, ARRL membership is relevant and important. There would be no Amateur Radio as we know it today were it not for the ARRL. We would be happy to welcome you as a member! (An Amateur Radio license is not required for Associate Membership.) For more information about ARRL and answers to any questions you may have about Amateur Radio, write or call:

ARRL—The national association for Amateur Radio
225 Main Street
Newington CT 06111-1494
(860) 594-0200
Prospective new amateurs call:
800-32-NEW HAM (800-326-3942)
You can also contact us via e-mail at **ead@arrl.org** or check out *ARRLWeb* at **http://www.arrl.org/**

Please use this form to give us your comments on this book and what you'd like to see in future editions, or e-mail us at **pubsfdbk@arrl.org** (publications feedback). If you use e-mail, please include your name, call, e-mail address and the book title, edition and printing in the body of your message. Also indicate whether or not you are an ARRL member.

On what other Amateur Radio subjects would you like the ARRL to publish books?

Where did you purchase this book? ☐ From ARRL directly ☐ From an ARRL dealer

Is there a dealer who carries ARRL publications within:

☐ 5 miles ☐ 15 miles ☐ 30 miles of your location? ☐ Not sure.

License class:

☐ Novice ☐ Technician ☐ Technician Plus ☐ General ☐ Advanced ☐ Amateur Extra

---

Name ARRL member? ☐ Yes ☐ No

______________________ Call Sign ______________

Daytime Phone ( ) ______________________ Age ______________

Address ______________________________________

City, State/Province, ZIP/Postal Code ______________________

If licensed, how long? ______________ e-mail address ______________

Other hobbies ______________________

____

Occupation ______________________

For ARRL use only	VAC
Edition	1 2 3 4 5 6 7 8 9 10 11
Printing	3 4 5 6 7 8 9 10 11

From ______________________________

____

____

Please affix postage. Post Office will not deliver without postage.

EDITOR, VERTICAL ANTENNA CLASSICS
ARRL
225 MAIN STREET
NEWINGTON CT 06111-1494

please fold and tape

VERTICAL
ANTENNA
CLASSICS
PROOF OF
PURCHASE